Advances in

Oilseed Research

Advances in Oilseed Research

R.K. Bhardwaj

Oxford Book Company

Jaipur ▪ New Delhi

Published By

Oxford Book Company

Regd. Off. "Bony Residency", Gate No. 2
Opp. Tilak Public School, Vishweswariya Nagar,
Gopalpura Road, Jaipur - 302018, India
Ph: 0141 - 2761280, Telefax: 2761381
email - oxfordbookjpr@gmail.com
website - oxfordbookcompany.com

Delhi Office: 102, 1st Floor, Satyam House,
4327/3, Ansari Road, Darya Ganj,
New Delhi - 110002, India
Ph: 011-45652440

Distributed By

KEYA INTERNATIONAL PRESS LLP
E-mail: keyainternationalpressllp@gmail.com

Published in India, 2024

ISBN : 9788189473815

Price: ₹ 1880.00

Typeset by Shivangi Computers, Jaipur

Printed and bound in India by Thomson Press Ltd., New Delhi

Preface

Used for a variety of reasons ranging from automobile field to cooking, oils are generally of three types depending upon their source: mineral, animal and vegetable oil of these the last form of oil is not only the cheapest, it is also the most viable form of oil, since it is renewable. As oil resources deplete throughout the world (especially in case of mineral oil), scientists have been hard pressed to search for more reliable source of oil production. Oilseeds are thereby a major option which scientists are continually exploring, such as linseed and poppy seed, and continuous research and development in the field is necessary for various uses.

The present text has been designed for scientists in the field of oilseed research, as a manual intending to keep them abreast of the latest trends and developments in the science. It brings within its purview the whole idea behind the research process, taking into current significant developments, the principles, practices and techniques behind them. In addition, it also incorporates an in-depth discussion of the progress made in various areas through oilseed research and the critical perspectives offer an insight into the future.

R.K. Bhardwaj

Contents

1

Oilseed Crops

Oilseed crops have been the backbone of agricultural economy of India from time immemorial. Today these crops are cultivated on about 16.5 million hectares, with total production of 10 million tonnes. This area constitutes approximately one-tenth of the total cultivated area in India. On the oilseed map of the world, India occupies a prominent position, both in reagard to acreage and production. The important oilseed crops grown in this country in order of importance are groundnut, rapseed and mustard, sesame, linseed, safflower, castor, sunflower and niger.

GROUNDNUT

Geographical origin: Groundnut (Arachis hypogaea L.) is believed to be the native of Brazil to Peru, Argentina and Ghana, from where it was introduced into Jamaica, Cuba and other West Indies islands. The plant was introduced by Portuguese into Africa from where it was introduced into North America. It was introduced into India during the first half of the sixteenth century from one of the Pacific islands of China, where it was introduced earlier from either central America or South America.

Economic importance: The oil content of the seed varies from 44 to 50 per cent, depending on the varieties and agronomic conditions. Groundnut oil is an edible oil. It finds extensive use as a cooking medium both as refined oil and Vanaspati Ghee. It is also used in soap making, and manufactoring cosmectics and lubricants, olein stearin and their salts. Kernels are also eaten raw, roasted or sweetened. They are rich in protein and vitamins A, B and some members of B_2 group. Their calorific value is 349 per 100 grammes. The H.P.S. type of groundnut kernels are exported to foreign countries.

The residual oilcake contains 7 to 8 per cent of N, 1.5 per cent of P_2O_5 and 1.2 per cent of K_2O and is used as a fertiliser. It is an important protein supplement in cattle and poultry rations. It is also consumed as confectionary product. The cake can be used for manufacturing artificial fibre. The haulms (plant stalks) are fed (green, dried or silaged) to livestock. Groundnut shell is used as fuel for manufacturing coarse boards, corksubstitutes etc. Groundnut is also of value as rotation crop. Being a legume with root nodules, it can synthesise atmospheric nitrogen and therefore improve soil fertility.

Botanical description: Groundnut (Archis hypogaea L.) is a member of sub-family, Papilionaceae of the family Leguminosae. Archis hypogaea L. consists of two subspecies each containing two botanical varieties.

— Subspecies hypogaea
  — variety hypogaea (The Virginia Group)
  — variety hirsuta Kohlar
— Subspecies fastigiata Waldron
  — variety fastigiata (the Valencia Group)
  — variety vulgaris (the Spanish Group)

Plants of the botanical variety hypogaea are spreading (runner) to upright (erect bunch) in growth habit, have alternate branching, lack inflorescences on the main stem, possess appreciable fresh seed dormancy, flowers are longer and mature later than those of subspecies fasligiata. Variety hirsuta has been used only to a little extent. Plants of this subspecies fasligiata are upright, have sequential branching and inflorescences in the main-stem leaf axils, possess little fresh seed dormancy, and are of shorter duration than those of the subspecies hypogaea. Subspecies fasligiata includes both the Spanish and Valencia types

Groundnut, in general, has a short-statured plant, with the main axis being upright (15 to 40 cm long) but the major part of the plant consists of the primary branches. Secondary and tertiary branches are found in the semi-spreading and spreading (Virginia) types, giving them a prostrate stature. The leaves are alternate, stipulate and quadri-foliate.

The flowers are orange yellow, typically papilionaceous, with a long calyx tube within which is held the style borne on a superior ovary. The calyx and corolla lobes are borne in the axils of the leaves on the fruiting branches. They are bisexual, Zygomorphic, complete and sessile. Petals are five, with one large standard, two wings and two fused keel petals. There are two steering and eight fertile anthers, four of which are globose and four oblonge (dimorphic).

Groundnut is predominantly a self-pollinated crop and pollination takes place early in the morning. As soon as the fertilisation is complete, the flowers fade. After fertilisation, an intercalary meristem becomes active at the base of the ovary above the point of attachment of the hypanthium, producing new tissue below itself and resulting in an elongated stalk in the peg.

The pegs are positively geotropic, enter the soil and bend in a horisontal plane. Generally two, and occassionally one, three or four, fertilised ovules are borne at the tip of the peg which later swells to become the pod. The testa is generally pink, but varieties with red, white, purple and blotched testa, with various gradations of colours are available.

Distribution and Production

The major groundnut-producing countries of the world are India, China, Nigeria, Senegal, Sudan, Burma and the USA. Out of the total area of 18.9 million hectares and the total production of 17.8 million tonnes in the world, these countries account for 69% of the area and 70% of the production. India occupies the position, both in regard to the area and the production, in the world.

About 7.5 million hectares is put under it annually and the production is about 6 million tonnes. 70% of the area and 75% of the production are concentrated in the four states of Gujarat, Andhra Pradesh, Tamil Nadu and Karnataka. Andhra Pradesh, Tamil Nadu, Karnataka and Orissa have irrigated area forms about 6% of the total groundnut area in India.

Climate and Soil

Groundnut is grown throughout the tropics and its cultivation is extended to the subtropical countries lying between 45 degrees N and 35 degrees S and upto an altitude of 1000 metres. The crop can be grown successfully in places receiving a minimum rainfall of 1,250 mm. The rainfall should be well distributed well during the flowering and pegging of the crop.

The total amount required for presowing operations is 100 mm; for sowing, it is 150 mm and for flowering and pod development an evenly distributed rainfall of 400-500 mm is required. The groundnut crop, however, cannot stand frost, long and severe draught or water stagnation. Groundnut is grown on wide variety of soil types. However, the crop does best on sandy loam and loamy soils and in the black soils with good drainage. Heavy and stiff clays are unsuitable for groundnut cultivation as the pod development is hampered in these soils.

Mixed Cropping

Generally, as a kharif crop, groundnut is grown year after year. In certain places, it is rotated with wheat, jowar, bajra, garden crops, such as potatoes, onions, chillies, garlic, ginger and turmeric. The yields of the cereal crops following groundnut are usually increased by about 25%. Pulses, e.g. red gram (arhar), mash and moong are grown mixed with groundnut. In certain places, millets e.g. jowar and bajra, and castor are grown mixed with groundnut.

Season

Groundnut is raised mostly as a rainfed kharif crop, being sown from May to June, Depending on the monsoon rains. In some areas or where the monsoon is delayed, it is sown as late as August or early September. As an irrigated crop it is grown to limited between January and March and between May and July.

Cultivation

For a kharif crop, with one set of rains in May-June, the field is given two ploughings and the soil is pulvarised well to obtain a good tilth. The third ploughing may be given just before sowing. Harrows or tillers can be used for cultivation. When the soils are heavily infested with perennial weeds, e.g. Cynodon or Cyperus, very deep ploughing is needed. If a field is infested with white grubs, chemicals, such as Heptachl or chlordane, are drilled at the rate of 25 kg per ha before final horrowing.

For the irrigated crop, beds of convenient size may be made, depending on the tropography of the land, the nature of the irrigation source and the mode of liftinfg water. Since groundnut is a deep rooted plant it uses up both moisture and nutrients in the deeper layers of the soil. Generally, the following fertiliser schedules are recommended, depending upon the soil fertility, the variety grown, and the quantum and distribution of rainfall.

Farmyard manure		*N*	*P*	*K*	
Rainfed crop	6.25 Tonnes	10-20	20-40	20-40	kg/ha
Irrigated crop	12.5 Tonnes	20-40	40-90	20-40	kg/ha

Usually, the application of only N and P is resorted to, since most of the Indian soils are rich in K. The application of N in two equal split doses, one before sowing and the other 30 days after sowing, is found to increase the yield. In virgin lands, when groundnut is newly introduced, the application of a culture of *Rhizobium* as seed treatment is beneficial in increasing nodulation and nitrogen fixation. The application of gypsom at 500 kg per ha at the pegging stage will enhance pod formation.

Healthy and well developed pods should be hand shelledor shelled with a suitable groundnut_sheeer about fornight before sowing. Well defilled kernels should be selected and treated with 5 gm of thiram or 3 g of Captanper kg of kernels. The kharif crop is sown with a speed-drill or with a suitable planter at a depth of 8 to 10 cm. specings adopted differ from place to place. For the rainfed bunch groundnut the inter-raw specings vary between the adjacent rows varies from 20 and 30 and intra-raw specings between 10 and 20 cm.

The quantity of well developed seeds required per hectare with the above specings will be about 110 kg for semi-spreading and spreading varieties and 120 kg for the bunch varieties. Weeds cause considerable reduction in yield. A reduction of 20 to 45% in yield due to weeds has been recorded. For controlling weeds, and also to keep the soil in a friable condition, the crop should generally receive a hand-weeding and one or two hoeings, with bullock-drawn implements, the first about three weeks after sowing and the second and the third about a fortnight and a month later. No interculture would be done after the pegs have commmenced going underground.

Earthing up can be done in the case of bunch and semi-spreading types to facilitate the maximum penetration of the pegs into the soil. Weeding can also be done quickly and efficiantly with a wheel hand-hoe, if the soil is light. It should be run before weeds make excessive growth. Weeds can also be controlled effectively with Lasso or Tok-E-25 weedicides at the rate of 5 litres in 500 litres of water per hectare as a pre-emergence soil spray within two days of sowing groundnut.

If the Kharif crop is cought in a long spell of draught, especially at the pod formation stage, supplemental irrigation is given, if water is available. For the irrigated groundnut, the frequency of irrigation depends on soil texture, and interval between irrigation ranges fro 8 to 12 days. Studies on the water requirement of indicate that the period between the fifty first and fifty seventh day the crop depends on timely irrigation to it during this period. The prominant symptoms of maturity are the yellowing of leaves, the shedding of older leaves, the development of proper colour of the testa and the dark tint inside the shell. The bunch and semispredading varieties are usually harvested by hand pulling when there is adequate moisture in a soil.

The spreading types, on the other hand, are harvested by digging with a mammity or by ploughing or working a blade harrow. The pulled out plants are stacked for a few days for drying and stripped afterwards. The stripped pods are cleaned and dried to a safe moisture content of not more than 5%. Damp nuts, if stored, will ferment and allow the development of poisonous moulds, e.g. aspergillus flavus in the Kernels, leading to contamination with aflatoxin—a health hazard both or human beings and livestock.

The oil express from such produced will be rancid and the cake. When fed to poultry results in a heavy mortality of the birds. It is desirable to store groundnut in gunnt bags are stacked in a storeroom in tires comprising not more in each in such a way that the air keeps circulating planks to avoid damage from dampness, rats, etc. The store rooms should be periodically inspected to ensure that there is no storage pest or to decompose of the stock immediately after proper cleaning.

Pest and Diseases

Aphids (Aphis craccivora Koch-), Leaf miner (Stomopteryx subseciva Zell)., the red hairy cater piller (Amsacta albistriga Walk., and A moorei Bltr.) and the white grub (Lachnosterna consanguiea) are the serious pests of the groundnut crop. Dusting B.H.C 10% on the young larve or spraying Metasystox) 0.05% on grown up catrepillers is recomended for controlling the red hairy caterpiller.

Spot-treatment with Phorate or Mephospholan granules at the rate of 1.5 kg a.i. per ha into the soil before planting is recomended for controlling the white grub. The tikka leaf-spot (Cercospora arachidicola and C. personata) and the coller- rot are important diseases of groundnut. Rust (Puccinia arachidis) has been occurring in a serious form in recent years in certain groundnut growing areas. It is desirable to store groundnut in gunnt bags are stacked in a storeroom in tires comprising not more in each in such a way that the air keeps circulating planks. Dry seed dressing with Thiram or captain @ 4g/kg of seed is recomended for controlling collar-rot. Yet spraying with Hinosan at the rate of 0.02% is recommended to check the spread of the Pathogen.

Yield

Under rainfed conditions, the average yield of semi-spreading and spreading varieties is 1200-1400 kg of unshelled pods per hectare and that of bunch types is 800-1,000 kg. The crops grown with supplemental irrigations produce 3,000 kg more of pods per hectare. The pods yield 70 to 75 per cent of kernels by weight. The yield of haulms is usually two to two-and-half times that of pods.

Varieties

The varieties under cultivation fall into three groups in respect of the habit of growth, namely bunch (Spanish), semi-spreading (Virginia bunch) and spreading (Virginia runner). In the bunch group, the plants grow erect, possess light-green foliage, produce pods in clusters at the base of the plant and have round, plump non-dormant seeds, with light-rose testa.

In the case of the semi-spreading and spreading varieties, the branches trail either partially or completely on the surface of the soil, produce pods all along them, possess dark-green foliage and have oblong, dormant brownish seeds. The semi-spreading and spreading types are usually heavier yielding and later-maturing than the bunch varieties.

RAPESEED AND MUSTARD

Under the names rapeseed and mustard, several oilseeds belonging to the crucieferae are grown in India.

1. Brown mustard, commonly called rai (raya or laha)—Brassica juncea (L.) Czern. & Coss
2. Sarson
   - Yellow sarson—B. campestris L. var. sarson Prain
   - Brown sarson—B. campestris L. var. dichotoma Watt
3. Toria (lahi or Maghi Labi)—B. campestris L. var. toria Duth.
4. Taramira or tara (eruca sativa Mill.)</LI< ol>

In trade, sarson, toria and Taramira are known as rapeseed, and rai as mustard.

Banarsi rai (B. nigra Koch.) which does not fall under any of the four groups is a garden crop used as spice. The cultivation of white mustard (Brassica alba syn. Sinapis alba) is no longer found in India. Rai and Yellow sarson are self-fertile and the rest of the cruciferous oilseeds, viz. brown sarson, toria, taramira, Banarasi rai and white mustard, are self-incompatible. The acreage under yellow sarson is scantily and constantly on the decrease. In the recent past, the accreage under brown mustard is steadily on the increase at the expense of other Brassica due to its higher production, greater to pests and diseases and moisture stress.

Origin

Brassica juncea L. (rai) was originally introduced from China into north-eastern India. From where it has extended into Afghanistan via the Punjab. Eastern Afghanistan, together with the adjoining north-western India is one of the independent centres of brown sarson (Brassica campestris var. brown sarson). Yellow sarson (B. campestris var. yellow sarson) is commonly grown in the eastern parts of India where it shows much diversity of forms. Taramira is a relatively recent introduction into India. It is believed to be a native of southern Europe and North Africa.

Economic Importance

Rapeseed and mustard yield the most important edible oil content of the seeds of different ranges from 30 to 48 per cent. In the case of white mustard, the oil content ranges from 25 to 33 per cent. The oil obtained is the main cooking medium in Northern India and can not be replaced by any other edible oil. The seed and oil are used as a condiment in the preparation of pickles and for flavouring curries and vegetables. The oilcake is mostly used as cattle feed. The leaves of young plants are used as a green vegetable. The use of mustard oil for industrial purposes is rather limited on account of its high cost.

Botanical Description

Rapseeed and mustard include annual herbs ranging in height from 0.45 to 1.75 m. Roots, in general, are long and tapering. Toria is more or less a surface-feeder and brown sarson has long roots, with a limited lateral spread, enabling its successful cultivation under drier conditions. Yellow sarson has both extensive and lateral spread. The height of the stem varies from 45 cm (in some varieties of toria) to 1.75 m (in yellow sarson). In toria and brown sarson, the branches arise at an angle of 30° to 40°.

In yellow sarson, branches arise laterally at an angle about 10° to 20° and give the plant a narrow and pyramidal shape. The inflorence is coroymbose raceme. In the case of sarson the four petals are spread apart, whereas in brown sarson and toria the petals overlap or may be placed apart, depending upon the variety. The flowers bear a hypogynous syncarpous ovary. In brown sarson and toria, the ovary is bicarpellary, whereas in the case of yellow sarson, it may also be tri or tetra-carpellary. The pods are two-valved, three-valved or four-valved, depending upon the number of carpels in the ovary. The flowers begin to open from 8 a.m. and continue up to 12 noon.

Distribution and Production

The crop is grown both in subtropical and tropical countries. In Asia, it is chiefly grown in China, India and Pakistan. It is also grown in Europe, Canada and the USSR, but the forms of rapeseed and mustard grown there are different from those grown in India. India occupies the first position, both with regard to acreage and production of rapeseed and mustard in the world. In India the Braddica crops occupy the second largest position after groundnut, with 3-5 million hectares, producing about 2 million tonnes of seed annually. The chief states producing them are Uttar Pradesh, Punjab, Haryana, Assam, Bihar, Madhya Pradesh, Rajasthan, West Bengal and Orissa.

Climate and Soil

The rapeseed and mustard crops are of the tropical as well as of the temperate zones and require relatively cool temperatures for satisfactory growth. In India, they are grown in the rabi season from September-October to February-March. The rapeseed and mustard crops grow well in areas having 25 to 40 cm of rainfall. Sarson and toria are preferred in low-rainfall areas, whereas raya and toria are grown in medium and high-rainfall areas respectively. The rapesed and mustard thrive best in light to heavy loams. Raya may be grown on all types of soil, but toria does best in loam to heavy loam. Sarson is suited to light-loam soil and taramira is mostly grown on very light soils.

Cultivation

A fine seed-bed is required to ensure good germination. In irrigated areas, the first ploughing is done with a medium sized soil-turning plough, followed by two to four ploughing with a desi plough or a cultivator. Sohaga (Planking) is given after every ploughing. In rainfed areas, one to two ploughings with a desi plough or a cultivator, each followed by planking, may be given.

Toria, in particular, requires a fairly moist seed-bed for good germination, but excessive moisture should be avoided. The rate in the case of mixed cropping depends on the proportion of the rapeseed to the main crop. When sown pure, 5 kg of seed per ha is used for all rapeseeds and mustard. When sown mixed with other crops, the time of sowing rapeseed and mustard is governed by the sowing of the main crop. The first half of September is best for sowing toria, 25th September to 15th October for sarson, 30th September to 15th October for raya, and tarantira is sown throughout October.

The seed from healthy and desirable plants, grown in isolation in the case of self-sterile form, should be used. Whenever moisture in the field is inadequate, the seed is mixed with moist soil and kept overnight. For distributing evenly, the seed is usually mixed with sand before sowing. In mixed cropping rapeseed and mustard are sown in rows 1.8 to 2.4 metres apart across the main crop. The pure crop of rapeseed or mustard is sown at a depth of 4 to 5 cm in lines, 30 cm apart, with a drill, or with a (tube) attached to the plough. Thinning is done three weeks after sowing to maintain a plant-to-plant distance of 10 to 15 cm.

One hoeing in the case of toria in the third week after sowing and 1-2 hoeings in the case of sarson and raya are adequate. Forty kg of W per ha is optimum for all rapeseed and mustard crops in rainfed areas. Under irrigated conditions doses of 40 and 80 kg of W per ha are considered optimum for toria, sarson and rapeseed respectively. For tarantira, 20 kg of W per ha is ancient. All the fertilisers should be drilled before sowing. Among the Brassicae, raya is most responsive to irrigation, followed by yellow sarson.

Two irrigations, one at flowering and the other at pod formation, result in the maximum yield in the case of toria, sarson and raya. Its flowering is most economical, if there is water scarcity. Tarantira is not irrigated. Harvesting is done as soon as the crop begins to turn yellow. Tma, which takes 75 to 90 days to mature, is the earliest crop to be harvested. Harvesting is done with hand-sickler.

Threshing is done by beating with a wooden stick the seed bearing part of the plants, taken in convenient bundles or by trampling them under the feet. Winnowing is done with by slowly dropping the threshed produce from a basket held shoulder-high. The seed after being dried in the sun is stored in gunny bags or bins.

Pests and Diseases

The most serious pests of rapeseed and mustard are the mustard aphid (Lipaphis erysinbi (Holt) syn. Rhopalosiphun pseudobrwsicae (Davis), mustard sawfly (Athalia proximo King) and cutworm (Agrotis ypsi" Poott.). To control aphidy the spraying of Methyl Demeton 0.02% or limetboate 0.03% or Phosphomidon 0.0V/ two or three times, depending on the intensity is recommended.

The mustard sawfly is controlled by dusting 10% BHC at the rate of 25 kg per ha. Cutworms are controlled by drilling into the soil 5% dust of Aldrin or Heptachlor or Chlordane at the rate of 25 kg per ha. Alternaria blight (Alternaria brassicae Bork.) is the most widespread and destructive disease of rapeseed and mustard, whereas rust (Cystopus candidus) is serious in certain areas. Spraying with Dithane M-45 or Difolaton at the rate of 11 kg per ha is recommended for controlling these diseases.

Orobanche, a phanerogamic parasite, has also been reported to be serious in Punjab and in certain parts of Uttar Flradesh. Hand-pulling of the plants of the parasite and a long rotation are recomended. Yield. The average yield of rapeseed and mustard is about 500 kg per ha. Under good crop management, toria, which is a short-duration crop, gives a yield of about 450 to 650 kg per ha, sarson 800 to 1,000 kg and rai 1,000 to 1,200 kg per hectare.

SESAMUM

Sesamum (Sesamum indicicum L.), also known as sesame, til and gingerly, is an important and ancient oil-yielding crop. Africa is considered to be the primary centre of origin of this crop because of the presence of its diverse wild species in that continent. India is the secondary centre of origin and another secondary centre is Japan.

Economic Importance

The sesamum seed is a rich source of edible oil. Its oil content generally varies from 46 to 52 per cent. Its grains may be eaten fried, mixed with sugar or in the form of sweetmeats. Sesamum oil is used as a cooking-oil in southern India. It is also used for anointing the body, for manufacturing perfumed oils and for medicinal purposes. Sesamum-cake is a rich source of protein, carbohydrates and mineral nutrients, such as calcium and phosphorus. The cake is edible and is eaten avidly by working classes. It is also a valuable and nutritious feed for milch cattle.

Botanical Description

The genus Sesamum belongs to the family Pedaliaceae which has a wide distribution, covering tropical Africa, Madagascar, Arabia, India, Sri Lanka, tropical Australia and a few of the eastern islands of the Malayan Archipelago.

Sesamum is an erect and annual herb. The plant is simple or branched. The leaves are opposite below and alternate above. The flowers are zygomorphic, solitary, occasionally two or three together, auxiliary, short-pedicelled, borne on the upper part of the stem or branches geniculate. The tube, bent downwords and is dialated above the oblique base. The calyx is small, and five sectioned, 0.5-0.6 cm long. The corolla is tubular, campanulate, approximately 3 cm long, with an applicate margin and slightly bilabiate, the upper lip is entire and the lower one is tripartite.

There are four functional stamens and often one is sterile, didynamous. The antliers are sagittate. The ovary is superior, bilocular, but may be subdivided by false internal walls. The fruit is a capsule, erect and oblong. The capsule contains numerous small ovate seeds. The testa may be smooth or reticulate and may be white, yellow, reddish-brown or black. The seed is dicotyledonous, albuminous and oleaginous.

Distribution and Production

India, China Sudan, Mexico, Turkey, Burma and Pakistan are the important sesamum producing countries. India ranks first, both in the area and production of sesamum in the world. The annual area put under it in India is about 2-5 million hectares (45 per cent of the world hectarage) and the total production is nearly 52 thousand tonnes.

Sesamum is grown on 21 lakh hectares in only eight states, viz. Uttar Pradesh (673,000), Rajasthan (562,000), Madhya Pradesh (345,000), Andhra Pradesh (237,000) Maharashtra (139,000), Gujarat (118,000), Tamil Nadu (117,000) and Orissa (103,000. Among other states only Karnataka has a sizable area (68,000) under sesamum. In the remaining states it is grown only on a small area and hence is a very minor crop there.

Climate and Soil

Sesamum grows in the plains and at elevations up to 1,200 m. It cannot stand frost, continuous heavy rain or prolonged drought. It is grown in sandy-loam to heavy black soils. Most of the crop is confined to lighter soils and its cultivation in heavy soils is limited to certain regions. The pH of sesamum growing soils ranges from 5.5 to 8.2. The sesamum growing soils are shallow to medium in Andhra Pradesh, Orissa, Andhra Pradesh and Tamil Nadu and are deep in Rajasthan, Gujarat and Maharashtra.

Mixed Cropping

Sesamum is grown both as pure and mixed with other crops, whereas the semi-rabi and summer crops are taken as pure. The common component crops are red-gram (arhar) in Andhra Pradesh, bajra and red-gram in Gujarat, red-gram and jowar in Madhya Pradesh. cotton, groundnut and red-gram in Maharashtra, and black-gram, green-gram and horse-gram in Orissa.

Cultivation

In India, seasmum is grown in three seasons, viz. kharif, semi-rabi and summer. The kharif crop occupies over 70 per cent of the area, whereas the semi-rabi and summer crops occupy 20 per cent and 10 per cent area respectively. The kharif sesamum is sown in June-July with the onset of the monsoon and is harvested in December-January. The kharif and semi-rabi crops are entirely rainfed, whereas summer crop is grown under irrigation. The yield of the kharif crop is poor, whereas those of the semi-rabi and summer crops are high, as they are grown in rich soils and under better management.

The preparatory cultivation for the kharif crop is usually not thorough. For the semi-rabi and summer crops, the land is ploughed and harrowed repeatedly to secure a clean and fine seedbed. Except in Maharashtra and Gujarat, where line-sowing is practised, in all other states it is sown broadcast. The recommended spacings for the line-sown crop in different regions, however, range from 25 to 35 cm between the rows and 10 to 20 cm between the plants in the row.

The seed being small, it is often mixed with sand to ensure its even distribution and the drill is operated rather shallow to avoid deep sowing. After sowing, the seed is covered lightly with a brush harrow. The seed-rate varies from 3 to 5 kg per hectare. The seed should be treated with Captain or Thiram at the rate of 3 g per kg of seed to control seed-borne diseases. The rainfed crop is weeded and hoed once or twice and the irrigated crop is weeded often. Wherever the weed control with mechanical means is a problem, chemical weed control with pre-emergence application of Lasso at the rate of 3 litres/ha is recommended. The crop is generally recommended for different regions as 25-30 kg of W, 20-30 kg of P_2O_5 and 0-20 kg of K_2O per hectare. The crop is harvested when the leaves, stems and capsules begin to turn yellow and the lower leaves start shedding. To prevent the shedding of grains, the crop should not be allowed to become a drip in the field. The ripe plants are cut, carried to the threshing-yard, stacked for a week in the sun with the cut-ends downwards and well shaken or beaten to take out the grains from the dry capsules. Winnowing and cleaning completes the process.

Pests and Diseases

The leaf-eating caterpillar (Antigastra Catalaulmlis Dup.) and the gall-fly (Asphon- dylia sesalili Felt) are the serious pests of sesamum. Two disulphur (0.5 kg/ha) sprays of Carbaryl 0.2%, or N on the 40th day and the 60th day, however give an elective control of both the pests. Stem and root-rot (Macrophominia phaseoli Maubl.) Ashby, phyllody (virus, mycoplasina), bacterial leafspot (Pseudomonas sesami Matkoff) and leaf-curl are the important diseases of sesamum.

Seed treatment with Ciptan at the rate of 3-9 Per kg of seed for controlling Macfphomila rot, 3 sprays of Agrimyein at the rate 01-100 ppm at 15-day intervals for controlling bacterial leaf spot are recommended. Both phyllody and leaf-curl can be controlled by spraying, the insecticides, such as Phosphomidon 0.05% or Dimethoate 0.05% or Dichlorovas 0.05%, which reduce the population of the insect vectors. The average grain yield of the kharif crop grown unmixed is 200 to 500 kg per hectare and the average yields of the semi-rabi crop and of the irrigated summer crop are 300 to 600 kg per hectare.

LINSEED

Linseed (Linum uaitdissimum L.) belongs to the family

Linaceae and the genus Linum which has 100 species. It has been grown from ancient times for fibre (flax) and for its seed which is rich in oil. The oldest regions of cultivated flax are reported to be in Asia and on the Mediterranean coast.

Economic Importance

Linseed occupies a greater importance among oilseeds; owing to its various uses and special qualities. In India, it is grown mainly for seeds, used for extracting oil. The oil content of the seed varies from 33 to 47 per cent. Linseed oil is an excellent drying oil used in manufacturing paint and varnishes, oilcloth, waterproof fabrics and linoleum and as an edible oil in some areas. Linseed-cake is a very good manure and animal feed. Linseed straw produces fibre of good quality. Linseed is used in making paper and plastics.

Area and Distribution

Linseed is extensively grown in the countries of the temperate zone as well as in those of the tropical zone. The major linseed-growing countries are Argentina, the USSR, India, the USA, Canada, Pakistan and Australia. India accounts for about 1.9 million hectares, with a seed production of 4.98 lakhs of tonnes and occupies the third rank among the linseed-producing countries.

Australia and Canada have the highest productivity of about 7 quintals per hectare, whereas India averages 255 kg per ha. This yield is the lowest in the world. In India, Madhya Pradesh leads in yield and acreage, followed by Uttar Pradesh and Maharashtra, Bihar, Rajasthan, Karnataka and West Bengal also grow linseed in large areas. Madhya Pradesh and Uttar Pradesh together contribute to the national linseed production to the extent of about 70 per cent.

Climate and Soil

Linseed is grown in the range of latitudes between the 10th and 65th parallels, both north and south. Its cultivation is confined to low elevations, but it can he successfully grown up to 770 metres. Areas with the annual rainfall ranging from 45-75 cm are best suited for its cultivation. The seed crop does well under moderate cold, but the fibre crop grows best in cool moist climates.

In India, the crop is grown in the rabi season from September-October to February-March. Linseed can be grown on different kinds of soils, except the sandy and badly drained heavy clays or clay loams. It does well on the deep clayey black soils of central and peninsular India and on the alluvium loams of the Indo-Gangetic plains.

Cultivation

Fields of linseed are prepared as for other crops. Frequent ploughings are given to retain moisture and to check weeds in the field during the rainy season and a fine weed-free seedbed is prepared for sowing. BHC 10% or Aldrin or Chlordane 5% dust at the rate of 26-30 kg per ha is mixed within soil during the last ploughing before sowing to protect the seedlings against white-ants and the early damage from cutworms.

The main season for sowing linseed is October to November, depending upon the cessation of the rains. The crop is grown either broadcast or is drilled in lines 20 to 30 cm apart: In the case of broadcasting, which includes uterasowing also, the seed-rate is kept at 40 kg per ha. For line-sowing, the seed-rate varies from 20 to 510 kg per ha, depending upon the seed size. Thin sowing gives a good number of tillers, but for fibre and weed, thick sowing is desirable.

The application of fertilisers and irrigation increases the yield of the crop. Under rainfed conditions, the fertiliser dose including 30 kg of N and 15 kg of P per ha is given. If irrigation is available, two to three irrigations are sufficient. Under irrigated conditions, the linseed crop requires 60 kg of K and 30 kg of P per ha. The deep placement of fertiliser at sowing in the case of the rainfed crop gives better results.

Under irrigated conditions, nitrogen is applied in two splits, half the dose as basal and the other half at the first irrigation, 40 to 50 days after sowing. The utera crop is fertilised at the rate of 10 kg of N per ha applied at the time of sowing. The rainfed crop generally requires no interculture. When the crop is irrigated, one weeding becomes essential.

Pests and Diseases

Linseed gall midge (Dasyneura lini Barnes) is the most wide spread and serious pest of linseed. The growing of early varieties and mixed cropping are reported to reduce the incidence of the pest. Linseed rust and wilt are the two most serious diseases. The growing of resistant varieties is recommended to control these diseases.

The linseed crop starts maturing by the mid of February, depending on winter spread and sowing time. Plants turn golden yellow when the crop is mature and ready for harvesting. Harvesting is done when the crop is dead ripe with a sickle or by uprooting the plants.

When the fibre is also desired along with seed, the harvesting of the crop is done at the stage of capsule maturity even when the crop is slight green. Linseed is threshed after it is completely dry. The crop is threshed either by beating it with a stick or by trampling it under the feet of the bullocks. Hand-beating is better to ge lien fibre, along with seed. Seed is separated from the chaff by winnowing and is stored in a dry place.

Yield

The average yield of a pure crop varies from 210 to 450 kg per ha of seed under rainfed cultivation. The crop in northern India generally gives higher yield than in central and peninsular India. The irrigated crop may yield 1,200 to 1,500 kg per ha. The percentage of oil in the seeds varies from 37 to 43, but the commercial extraction in expellers of rotaries gives about 33 per cent from small seeds and 34 to 36 per cent from hold seeds. Crushed in village ghanis, the seeds yield only 25 to 30 per cent of oil.

CASTOR

According to the available literature, castor (Ricinus comrnunis L.) is indigenous to Eastern Africa and most probably originated in Ethopia.

Economic Importance

It is generally grown for its, oil-yielding seeds. The oil content of the seeds varies from 35-58 per cent in different varieties, the average being about 47 per cent. Castor-oil is being used widely for various purposes. It is used as a lubricant in high-speed engines and aeroplanes, in the manufacture of soaps, transparent paper, printing-inks, varnishes, linoleum and plasticisers. It is also used for medicinal and lighting purposes. The cake is used as a manure and plant stalks as fuel or as thatching material or for preparing paper-pulp. In the silk-producing areas, leaves are fed to the silkworms.

Botanical Description

Castor belongs to the gellus Ricinus, its a member of the Euphorbiaccae which consists a vast number of plants mostly native to the tropics. The geiut; Ricinus is considered to be momotypic and R. cwnmunis as the only species which includes many polymorphic types.

The stem is erect, circular in section, partially, hollow, smooth, glabrous, with good branching. The stem is marked by a number of well-defined nodes, from each of which a leaf arises. The lower internodes are shorter and their length increases with the height. Leaves are alternate, large, palniate with 5-11 lobes, acuminate, margins notchedy serrate or indented. They are carried on long stout petioles. The inflorescences are borne terminally on the main and lateral branches.

Flowers are large, in terminal sub-panelled racemes, monoccious, apetalous, the upper portion of the raceme being occupied by the female flowers and the lower by the male flowers. The fruit is a roundish glaucous capsule, with three projecting sides covered with tough spines or smooth, three-loculed and three-seeded. Seeds are albuminous, anatropous, broad, oval, compressed with a marked carnnele and longitudinal raphe. The testa is thin, brittle, varying in colour and mottling. Below the testa is the thin tegmen, covering the whitish oily endospeim containing the embryo.

Distribution and Production

The main castor-growing countries are Brazil, India, USSR and Argentina. India ranks first in area (44,8000 hectares) but second in production (140,000 tonnes), Brazil being the, first. Andhra Pradesh (67.2%), Gujarat (12.7%) Karnataka (7.1%) and Orissa (5.8%) account for over 90% of the area and also production. In lighter soils, the land is repeatedly ploughed in summer, as and when the rains are received, and with the onset of monsoon rains the clods are crushed by working a country plough or harrow.

In clay loams, only harrowing is done with a blade-harrow two to three times to bring the soil into proper tilth. The seed is sown either in the plough furrow or with a seed-drill or by hand-dibbling. The seed-rate used for sowing behind the plough is 12 kg per hectare. For hand-dibbling, 8 to 10 kg of seed will be sufficient per hectare. Before sowing, castor seed should be treated with Thiram 3 g/kg of seed to prevent the attack of root-rot and Alternaria blight.

The spacing adopted generally ranges from 90-120 cm between rows and 45-60 cm between plants in the row. One or two weedings and hoeings with bullock-drawn implements are given and sometimes, the plants are earthed up. The fertiliser dose recommended E for castor is 40 N-40 P-20 K kg/ha. The entire dose of P and K and half of W is applied basally and the remaining half of W about 30 days after sowing, as and when the soil moisture is adequate.

The improved varieties cultivated at present are early maturing and take about 150 to 180 days. When one or two capsules in a bunch show signs of drying, the whole cluster is generally removed and stacked, covered and weighed in a corner of the field or in a pit. The harvesting of unripe capsules in this manner has an adverse effect on the oil content of the seed. It is preferable to collect the fruits, as and when they ripen. When the whole crop is gathered, it is dried in the sun for a few days and the threshiug is done either by beating the dried capsules with a stick or by treading them under the feet of bullocks. Winnowing is done in the usual manner.

Pests and Diseases

The castor semi-looper (Achow janaga L.) and the capsule-borer (Dicitocrocis punctiferalis) are the two most serious pests of castor. Dusting BHC 10% in early stages or spraying 0.44% Ekalax or 0.1% Carbaryl or 0.05% 1 arathion on the crop will give an effective control on these pests.

The seedling blight (Phytopiathora colocasic Racib. emend. Thomas & T.S. Ramkar and Alternaria blight Alternaria ricini (Yoshii) Hansford cause serious losses to this crop. The Sowing of the crop in low-lying and water-logged areas should be avoided to prevent the seedling blight from appearing. For alternaria blight, seed treatment with Thiram at the rate of 3 g/kg of seed is recommended.

Yield

The average yield of rainfed castor varies from 200 to 500 kg per hectare, that of the mixed crop from 100 to 200 kg and of the irrigated crop from 500 to 800 kg. In eleven years (from 1964-65 to 1974-75), the average total yield of castor in India has risen to 350 kg per ha.

Varieties: The varieties differ in the branching habit of the plants, colour of the stem and branches (red and green), the nature of capsules (smooth or shiny), duration (early or late) and the size of seed. The following are the principal improved varieties of castor recommended for cultivation in different states.

SAFFLOWER

The three centres of origin, as suggested by Vavilov for safflower (Carthamus tinctorius L.), are India, Afghanistan and Ethopia. Decandole was of the opinion that Arabia was the most probable centre of origin.

Economic Importance

The crop is now cultivated, primarily for its seeds which yield oil, though at one time it used to be grown for the extraction of a dye also. The seeds are edible and are eaten after roasting. Their oil content varies from 24 to 36 per cent, depending on the variety, soil, climate and other conditions. The cold-pressed oil is golden yellow and is used for culinary purposes, or for making soap.

The oil obtained by dry hot distillation is black and sticky and is used only for greasing well ropes and leather goods exposed to water. Safflower oil has also good drying properties and is, therefore, used in the manufacture of paints, varnishes and linoleum. It can be mixed with white paint without any after-yellowing effects. The cake, particularly from decorticated seed, is used as a concentrated cattle feed, and that from undecorticated seed is sometimes used as a manure.

Botanical Description

The safflower plant is highly branched, herbaceous, thistle-like annual, varying in height from 30 to 160 cm. It has many branches each terminating in a flower. The inflorescence is broad, flat or slightly curved and densely bristled owing to the presence of numerous floral tracts. The flowers are yellow to red, containing 20 individual ficrets, each of which produces a seed.

Distribution and Production

The important safflower growing countries, besides India, are the USA, Mexico, Ethopia, Spain, the USSR. and Australia. In India, it occupies 590,000 hectares with a production of nearly 130,000 tonnes. Over 98% of the area is concentrated in the states of Maharashtra (04.4%), Karnataka (26.0%) and Andhra Pradesh (8.0%).

Climate and Soil

The crop is grown during rabi primarily as a rainfed crop, but in some areas it is raised under irrigation. At all stages of growth of the crop, excessive rainfall or humidity increases damage from fungal diseases. Water-logging due to poor drainage or prolonged rains even for relatively short periods can cause substantial fall in seed-yield. Being drought-resistant, it is cultivated on all types of soil, including sandy soil, but it thrives best on water-retentive black soils and alluvial loams. The crop is fairly resistant to saline condition.

Cultivation

Saffiower is grown mostly as a mixture with wheat, barley, gram and rabi jowar, three rows of safflower being planted after every nine, 12 or more rows of the main crop. Sometimes, it is sown as border rows surrounding the crops of wheat, rabi, jowar, etc., because being spiny, it protects the main crop against cattle trespass. As a pure crop, it is rotated with wheat, gram and rabi jowar. Occasionally, it is grown as a second crop after any quick-maturing kharif crop, e.g. green-gram (moong), black-gram (mash), groundnut, coriander, etc.The number of ploughings should be as few as possible and be aimed at preserving the maximum amount of soil moisture. One or two ploughings with a country plough will be sufficient to remove the weeds and break the clods. Sowing is generally done in September and October. The seed-rate adopted varies from 5 to 12 kg per hectare, depending upon the soil fertility and the nature of the crop (pure or mixed). When grown in strips, a row spacing of 45 cm is adopted.

Plants too close tend to have thinner stems or wear a superficial root-system, with a reduced number of flowers. In general, the pure crop is given one or two weedings, combined with hoeing on the 20th and the 45th day after sowing. The application of 20-40 kg of N per hectare results in a substantial increase in yield. The mixed crop shares the preparatory tillage, manuring and cultivation given to the main crop.

Harvesting and Yield

The crop matures in 4 or 5 months after sowing. The ripe plants are either cut or pulled out, stacked for a few days to dry, threshed by cutting with sticks, and the resulting material is winnowed to obtain clean seeds. The average yield of the pure crop is 400-500 kg of seed per hectare, whereas that of a mixed crop is about 100 kg.

Pest and Diseases

The safflower aphid, thrips and the flower bud fly are the serious pests of the crop. Spraying the crop with 0.1% Fenithion, 0.03% Dimethoate and 0.07% Endosulfan is recommended for controlling aphids, thrips and budfly, respectively. Rust, alternaria leaf-spot and root-rot are the important diseases of safflower. The growing of resistant varieties is recommended to avoid damage from these diseases.

Varieties: The varieties under cultivation in different regions of the country are given in Table 1.

Table 1.

State	*Name of the variety*	*Duration in days*	*Special characteristics*
Karnataka	'A.I'	125	
		125	Yields 800-850kg/ha, oil content 30.8%
	'A-300'	125	
		125	Yields 750-800kg/ha, oil content 30.9%
Madhya Pradesh	'NO-7'	140	Yields 700-850kg/ha, oil content 30.3%
Maharashtra	'NO-62.8'	140	Yields 1,000-1,200 kg/ha, oil content 30.%
	'NAG-7'	140	Yields 1,000-1,200 kg/ha, oil content 30.%
	Tara	130	Yields 1,000-1,400 kg/ha, oil content 32.5%
Tamil Nadu	'K-1'	120	Yields 600-800 kg/ha, oil content 30.5%
Andhra Pradesh	'Manjira'		Yields 1200 kg/ha, oil content
	C 438	110	32%

SUNFLOWER

The cultivated sunflower (Helianthus annus L.) is a native of southern United States and Mexico, from where it was taken to Spain before the, middle of the sixteenth century. In the nineteenth century, the cultivation of sunflower as an oilaced crop began in the Soviet Union and the majority of the present day varieties grown all-over the world trace back their origin to the USSR.

The major sunflower-producing countries of the world are the Soviet Union, Argentina, Bulgaria, Pumania, Turkey and South America. Sunflower, as an oilseed crop, was introduced into India in 1969 and, according to the estimates of the trade and the Vanaspati Manufacturers Association, 3,88,000 hectares of sunflower was grown in India during 1975-76.

Economic Importance

Sunflower is mainly grown for its oil. The oil is used for culinary purposes, in preparation of vanaspati and in the manufacture of soaps and cosmetics. It is especially recommended for heart patients. Its cake is rich in protein and is used as a cattle and poultry feed.

Botanical Description

Sunflower belongs to the family Compositac and the genus Helianthus. The cultivated genotypes are characterised by a single stem terminating in a capitulum. Sunflower is protandrous, in which the male and female elements mature at different times. There appears to be a time-lag of 18-24 hours in the maturity of the male and female elements. So it is essentially a cross-pollinated plant, besides showing varying degrees of self-incompatibility.

Climatic and Soil Requirements

Sunflower is a day nuteral palnt can be grown successfully

in different seasons under varying conditions of day length, provided the temperature is favourable. Sunflower, basically a temperature plant, can also be grown commercially in tropical or sub-tropical conditions. It grows best with clearsky and occasionally rain shower during early stages. This crops requires a cool climate during germination and seedling growth and warm non-cloudy weather and high temperature during flowering to maturity.

Sunflower can tolerate temperature range of 8 to 30°C but the temperature below 16°C and above 40°C reduces yield and oil content. High temperature affects pollination and fertilisation, thus care should be taken to adjust the sowing date in such a way that flowering does not coincide with high temperature. The optimum temperature for best performance under controlled conditions is 27-28°C. Sunflower can be planted successfully when soil temperature exceeds 10°C, though germination begins at 3-5°C.

Young plants can withstand frost to -6°C. However, plants are frost sensitive from 6-8 leaf stage until flowering finishes. Late frost does not destroy the whole plant but destroys the fruiting head, leading to increased branching, resulting in small heads with many empty fruits. During flowering temperature should be between 18-20°C. Goods yields are obtained where the ruling temperature is 18-22°C showing that generally speaking the crop needs a warm climate.

Average minimum temperature below 17°C between flowering and maturity is necessary to ensure best oil quality. The crop is grown from 45°S to 55°N but greater production is between 20-50°S. The crop can be cultivated upto an altitude of 2500m but highest yield of oil is obtained below 1500m. Sunflower cannot tolerate frost and requires a frost free period of 120 days. It is resistant to drought but requires continuous availability of soil moisture for the best performance. Sunflower can thrive on a wide range of soils such as sandy loam, black soil and alluvial soil, but it performs better on sandy loam soil than on clay soil under similar management. However, good drainage is more important than soil type. Thus, fertile, well drained, medium textured soils having heavy water holding capacity are best suited for the cultivation of sunflower. Sunflower is highly susceptible to water logging. Avoid cultivation of sunflower on soils affected by salts and with poor drainage. The optimum pH requirement is 6.0 to 8.5.

Preparatory Tiliage

Sunflower requires a fine seedbed, free of weeds and clods, for proper germination and good crop stand to bury the residues or previous crop, ploughing with disc harrow followed by 2-3 cultivations and plankings may be required. fields vacated by potato may not require much seedbed preparation. However, adequate soil moisture is essential forgetting good germination.

Variety

MSFH8, PSFH-67,GKSFH-2002, Jwalamukhi, Mega-363 and NSFH-592 have been recommended. These hybrids are medium tall with dark green leaves. Their seeds are black in colour and mature in about 110-125 days mega-363 is a long duration hybrid and hence should not be selected for cotton areas. These hybrids have an average yield about 20 quintals per hectare with an oil content of 40.5 to 43.0 percent. In hybrids fresh seed is to be sown every time which adds to its cost of cultivation.

Sowing Time

Spring season is the best for getting higher and stable seed yield sunflower. Sunflower being a cross pollinated crop depends on honey bees for pollination and seed setting.

Availability of honey bees is abundant during spring season due to favourable. Planting during spring should be made after the soil temperature exceeds 10°C at 10cm depth at 8 a.m. Earlier planting at lower temperature increases time of flowering without much earliness. Best results are obtained when crop is sown in the first fortnight of January which produces higher seed yield than later planting. Sowing of short duration hybrids like GKSFH-2002, PSFH-67 and nsfh-592, however, can be extended upto second fortnight of January. Besides low yields, the late sowing may also face early mansoon rain at harvesting ans threshing causing difficulty in its management.

Seed Treatment

Use 5kg seed per hectare for hybrids to achieve the required plant stand. Low seed rate recommendation for hybrids is mainly because of smaller seed size and wider spacing for hybrids. Seed should be treated with Brassicol or Thiram @ 2 g per kg of seed to control seed and soil borne fungal disease.

Sowing

A spacing of 60x30cm for hybrids has been found to be an optimum. These spacings result in better development of head, more seed filling and less stem breakage.

Crust Breaking

Rain after sowing, results in crust formation, which hinders the proper emergence of seedlings. Thus it may be necessary to break the crust, which can be done with harrowing. Harrow as soon as possible after the crust has performed and if possible, before seedlings have emerged.

FERTILISER MANAGEMENT

Fields studies have shown that hybrid sunflower responds

well to nitrogen application upto 60 kg/ha. further increase in nitrogen results in lower seed and oil yield. Application phosphatic fertilisers @ 30kg P_2O_5/ha increase sunflower yield. sunflower responds to K application only on soils deficient in available potassium and it is recommended to be applied @ 30 kg K_2O/ha on such oils. Regarding application method, drilling the fertilisers besides the seed row has been found to be more beneficial for getting higher seed yield.

Application of fertilisers below the seed was found to be injurious to germinating seeds and reduced seed emergence count and seed yield. Entire dose of N, P and K should be applied at the same time of sowing. Prefer single superphosphate for P application as it also contains sulphur to which sunflower responds favourably. If single superphosphate is not available apply gypsum @ 25-30 kg/ ha at sowing time.

A suppllementary dose of nitrogen @ 25-30 kg/ha may be top dressed a month after sowing sunflower if the crop exhibits symptoms of nitrogen deficiency. Nitrogen deficiency appears as yellowing of the older leaves, which die prematurely. Leaves are also fewer and smaller. Plants with a severe deficiency are yellow and stuned, while young plants are of purplish colour. Plants are slender with longer internodes. The flower head is small with many unfilled seeds and the oil content is reduced.

Phosphorous deficiency results is delayed flowering, stuned growth and small leaves which prematurely die back from the leaf tip. Phosphorous deficiency is shown in small dark green leaves later turning pale with necrosis. Growth is restricted with short stiff stems, flowers are small, stuned malformed. Ripening is delayed, oil content and yield are reduced. Potassium deficient plants are squat with a proportionately greater reduction in stem length then in leaf size. Initially leaves are dark green but turn yellow at the edges and die except the lower leaves.

Potassium deficient plants have restricted growth with a rosette arrangement of the younger leaves, later the leaves develop dark brown necrotic patches.

Water Management

Sunflower can withstand moderate drought because of its deep root system. However, it responds favourably to irrigation and its seed yield can be increased subtantially by sustantially by sustaining adequate moisture insoil by timely irrigations besides improvements in oil content. Sunflower is highly moisture sensitive plant and its broad leaves exhibit wilting with slight moisture deficiency.

First irrigation to sunflower should be applied a month after sowing and interval during the subsequent months be reduced to 2-3 weeks during March and 8-10 days during the hots months of Apriland May. Irrigation should be stopped about 12-14 days before harvest. Sunflower is a tall growing plant with heavy head, therefore, irrigation at flowering and thereafter should be given on clam days to avoid lodging. The incidence of Sclerotinia stem-rot is more under irrigated conditions especially where the soil drainage is poor. Excessive and frequent irrigations on fine textured soils should be avoided.

Weed Management

Mechanical methods such as hoeing at 3-4 weeks after sowing followed by earthing up before flowering has been found to be effective in checking weeds. Use of tractor drawn implements before the crop attains 60-70 cm height has also been found effective for better efficiency of manual weeding an improved 3 tine wheel hand-hoe could also be used. Removal of weeds through mechnical/manual means means quite effective but expensive. Moreover, often timely labour is not available. The studies initiated to find effective herbicides for weed control in sunflower, revealed that pendimethalin as pre-emergence application gave effective weed kill.

Harvesting

The crop is ready for harvesting when the back of flower heads turn lemon yellow or yellowish brown near the stalk and the discs start drying. At this stage the seeds give a blakish look and are fooly ripe. The shadding of disc-flower residues from undamaged heads also indicate psysiological maturity. Promptly harvesting seed after psysiological maturity has many advantages; seed is least exposed to bird and rodent attack; the crop cycle is shorter resulting in more time to prepare for the next crop; reduces harvest seed losses; provides high quality seed and reduces exposures to bad weather.

The harvested heads are mildly dried and threshed by beating or trampling with tractors or sunflower threshers. The sunflower thershers are more efficient and economical. After threshing the produce should be thoroughly dried before storing otherwise in storage it gets affected by fungus and decomposes.

Allelopathic Effects of Sunflower

Researches show that sunflower depresses the growth and productivity of succeeding crops. The reason postulated are the allelochemicals being secreted by sunflower crop. Decomposing residues of sunflower could also bring about allelopathic effects. Phytotoxic compounds released into the soil rhizosphere have been observed to cause varying degree of allelopathy/autotoxicity in crop.

Field studies conducted in the department of Agronomy, Punjab Agricultural University, Ludhiana on the allelopathic effect of spring sunflower on succeeding Kharif crops revealed that maize, sorghum, cowpea, pearlmillet, pigeonpea, mungbean and sesamum suffered

41.7, 36.7, 27.6, 28.8, 23.0, 10.8 and 9 percent reduction in yield, respectively, as compared to these when grown after wheat. Studies in allelopathic effects of different plant parts of sunflower indicated that sunflower roots reduced pigeonpea, cotton and mungbean yields the most followed by roots plus shoots treatment. Sunflower shoots did not have any harmful effect on growth of these crops.

Bioassay studies involving use of aqueous sunflower extract, decomposing plant residue and soil from area in proximity of sunflower crop revealed inhibition of gemination and growth of cotton, pigeonpea, sunflower and mungbean. These studies further showed that effect of leaf extract/residue was more than roots. The allelopathic effect of sunflower plant residue persisted upto 24 days after incorporation and beyond this period the effect was rather meagre.

Chromatograhic and physico-chemical analysis of allelochemicals in sunflower parts showed the presence of chlorogenic and isochlorogenic acids in both stems and leaves besides these some unidentified phytotoxins were also observed. The concentration of these was more in leaves than in stems. Further studies revealed that total phenolics was more in leaves followed by in roots and shoots. Sunflower raised under moisture stress conditions had higher phenolics as compared to the raised without any moisture stress.

CUT WORM

It is polyphagous pest adn cuts off the tender stalks, near ground, of young crop plants during night. During day time the caterpillars remain hidden in the soil. It is more serious during March-april when sunflower follows potato. When flat sowing is practised, apply 5 liters of Dursban/ Radar 20 EC (Chlorpyriphos) per hectare before sowing. These insecticides should be mixed in 25 kg fine soil and broadcast uniformly after last ploughing but before planking. Sow crop on ridges to avoid cut worm damage in the early stages.

Tobacco Caterpillar (*Spodoptera litura*)

The polyphagus pest remain active throughout the year feeding on one or the other crop. The newly hatched caterpillar is pale-greenish with a black head and feeds on the outer epidermis of leaves. It poses a serous threat during April-May and defoliates the plants.

Cabbage Semi-looper (*Plusia orichalcea*)

This insects feeds on many crops including the sunflower. The newly hatched caterpillars defoliate the crop and cause severe damage to it during April-May.

Bihar hairy caterpillar (*Spilosoma obliqua*)

It is also called Bhabbu Kuta. This is highly polyphagous and voraciousfeeder. The nefarious pest causes severe damage and plants looke denuded. Newly hatched larva is bright brown and its body is covered with hair and legs are brown. At the older stage it is much darker and with dark brown hair of almost uniform length. these insects feed gregariously and should be collected along with the plant and buried deep in soil.

For the control of above mentioned pests, spray 1.125 L. Folithion/Sumithion 50 EC (fenotrothin) or 1.225 L. Thiodan 35 EC (endosulfan) or 500 ml Nuvan 100 EC (dichlorvos) in 250 litres of water per hectare. Repeat after two weeks, if necessary.

Head Borer (*Helicoverpa armigera*)

Head borer or American boll worm has become the noxious polyphagous pest in India adn also causes heavy losses in sunflower. Newly hatched larva is green yellow which changes its colour with every instar.The full grown caterpillar is brownish green having grey longitudinal streaks on it. The adults lay eggs on the flower buds. After hatching the larvae feed on developing seeds and make tunnles through them.

Head borer can be controlled by spraying 2.5 Kg of Sevin 50 WP (Carbaryl) or 2.0 litres of Asataf 75 SP (Acephate) or 2.125 litre Folothion 50 EC (Fenitrothion) or 2.5 litres of Dursban 20 EC (Chlorpyriphos) or Thiodan 35 EC (Endosulfan) or 1.25 litre of Nuvacron 30 SI (Monocrotophous) in 375 litres of water per hectare. Repeat after two weeks if necessary. Endosulfan is comparatively safe for honey bees. Spray when honey bee activity is minimum i.e. early morning or late evening.

NIGER

Geographic Origin: The Niger (Guizotia abyssinica L.F. Cass. crop is said to be indigenous to tropical Africa and more specifically to Ethiopia.

Economic Importance: The crop is grown for its seed, used for extracting oil which is about 37 to 43 per cent of the seed weight. The oil is used for culinary purposes, for anointing the body, for manufacturing paints and soft soaps, for lighting and lubrication and for manufacturing cosmetics. The oilcake is a well-known cattle feed. The crop sheds a large quantity of dry leaves in the field and thereby adds organic matter to the soil.

Botanical Description: Niger belongs to the family Compositae. The plant is a stout, erect, smooth orscabrous leafy annual herb, about one metre tall. The leaves are opposite or the upper ones are alternate. The capitulum is peduncled, auxiliary and terminal heterogamous. Bay-florets are pistillate, scriate, fertile; ligule 2-3 toothed and yellow. Disc florets are hermaphrodite, fertile; limb campanulate and 5-fid. Achenes are glabrous, dorsally compressed, tip rounded, pappus absent.

Distribution, area and Production: India is considered to be the chief Niger-producing country in the world, with an area of 4.8 lakh hectares, and with an annual estimated production of 1.2 lakh tonnes of seed. It is mainly grown in the states of Madhya Pradesh, Bihar, Maharashtra, Orissa and Tamil Nadu, of which the largest area of 2.38 lakh hectares is in Madhya Pradesh.

Climate and Soil: Niger is grown as a kharif crop under rainfed conditions. A moderate rainfall of 1,000-1,250 mm suits this crop and, as such, it is not cultivated in regions of heavy rainfall.

Light red soils and brownish loams with sufficient depth and good texture are considered suitable for this crop. It is often grown on poor soils of coarse texture. It can be grown on well-drained heavy soils.

Cultivation. Niger is grown either as a pure crop or is grown mixed with minor millets. Two to three ploughings are enough to prepare the land for sowing. The field is made weed-free and levelled by planking before sowing. The crop is sown from June to August, either behind a country-plough or with a four-coultered drill. A spacing of 30 cm between lines and that of 10-15 cm between plants in the line are generally adopted.

A seed-rate of 7-8 kg per hectare is adequate to maintain a good crop-stand and a pure crop. About a fortnight after sowing, thinning is done to regulate spacing between plants. One or two hoeings are required for controlling weeds. In general, the Niger crop is not manured. But the application of 20 kg of N and 20 kg of P_2O_5 per hectare is recommended for obtaining higher yields of the crop. When sown as a mixed crop, Niger shares the manuring given to the main crop. The crop matures in November and December. The mature plants are cut with sickles at the base and heaped in the threshing-yard for a week. They are then spread in the sun to dry for two to three days, threshed by beating with sticks, and winnowed.

REFERENCES

Saharan, G.S., et al., *Diseases Of Oilseed Crops, Indus Publishing Company,* ISBN: 8173871760.

David V. Alford, *Biocontrol of Oilseed Rape Pests,* 2003.

Chand, Ramesh, 2004, "WTO and Oilseeds Sector: Challenges of Trade Liberalization", *Economic and Political Weekly,* February 7, 533-537.

Dohlman, E., S. Persaud, and R. Landes, 2003, "India's Edible Oil Sector: Imports Fill Rising Demand", OCS-0903-01, *Electronic Outlook Report from the Economic Research Service,* USDA.

2

Oilseed Processing

Oilseed processing expands the use of crops such as sunflowers and brings value to grape seeds, which are usually waste products. Farmers and small-business owners are asking whether it is possible and profitable to add value to their seeds and nuts by extracting the oil. This is not an easy question to answer because there are so many variables. Most oil processing in the U.S. is done on a large industrial scale, and the processes use proprietary information.Small-scale oil extraction is more commonplace in other parts of the world, thus many of the useful resource materials and much of the appropriate-scale machinery come from other countries.

PROCESSING STEPS

Preparation of the Raw Material

Preparation of the raw material often includes removing husks or seed coats from the seeds and separating the seeds from the chaff. Information about seed cleaners designed for small-scale operations is often available from agricultural research stations doing small plot research with various crops. For instance, USDA researchers working with flax in North Dakota buy some of their equipment from Seedburo, which will supply a catalog upon request. Almaco is another company that specialises in equipment for research, such as one- and two-row test plot combines and various types of stationary threshers.

Oil Extraction

Oil can be extracted mechanically with an oil press, an expeller, or even with a wooden mortar and pestle—a traditional method that originated in India. Presses range from small, hand-driven models that an individual can build to powerdriven commercial presses. Expellers have a rotating screw inside a horizontal cylinder that is capped at one end. The screw forces the seeds or nuts through the cylinder, gradually increasing the pressure. The material is heated by friction and/or electric heaters.

The oil escapes from the cylinder through small holes or slots, and the press cake emerges from the end of the cylinder, once the cap is removed. Both the pressure and temperature can be adjusted for different kinds of feedstock. The ram press uses a piston inside a cage to crush the seed and force out the oil. Oils can also be extracted with solvents, but solvent extraction is a complex operation.

The enclosed Intermediate Technology Development Group Technical Brief, Principles of Oil Extraction, describes several methods of extraction. It cautions that "solvent extraction is not suitable for small-scale processing because of high capital and operating costs, the risk of fire and explosions from solvents, and the complexity of the operation." Waste management would probably be a problem, also.

Clarification

Clarification removes contaminants, such as fine pulp, water, and resins. You can clarify oil by allowing it to stand undisturbed for a few days and then removing the upper layer. If it needs further clarification, filter the oil through a fine filter cloth. Finally, you can heat the oil to drive off traces of water and destroy any bacteria.

Packaging and Storage

Use clean, dry containers to package and store oils and help prevent rancidity. Sealed glass or plastic bottles are adequate. Coloured containers in a dark box help to increase shelf life. The shelf life of oil is usually 6 to 12 months, if it is properly packaged and kept away from heat and sunlight.

SUNFLOWER SEED OIL

Ram Press

Enterprise Works Worldwide (EWW—formerly Appropriate Technology International) has supported a number of projects in Africa that process oilseeds using the ram press. The presses weigh about 45 pounds and can process 100 to 120 pounds of sunflower seeds per day. The extraction rate is 20 to 25% using the softer-shelled, high-oil-content sunflower varieties. The ram press can also be used for sesame, mustard, safflower seeds, and peanuts.

Choose seed that has a soft shell and high oil content (40–45% oil). Most high-oil-content sunflower seeds are all black. Many striped seeds have little oil. To get some idea of how the seed will work in the press, try to crack it open with your fingers. This will not be difficult with soft seed. With hard seed, you may need to use your teeth. Now look at the kernel inside the shell. It should fill the shell completely. If the seed is much smaller than the inside of the shell, it was probably harvested too early. You will not get much oil from it.

Put the kernel (without the shell) between your two thumbnails and squeeze. Did the kernel fall apart? Are there little smudges of oil on your thumbnails? If not, there may not be much oil in the kernel. (This is a very rough way of assessing seed. The manual also presents a more precise way to determine the percent of oil, but still encourages the practice of checking seeds with your fingers.) Check the different batches of seed you come across. In time you will get a feel for the differences.

For successful pressing, the seed must be:

— *Dry:* Moist seed will lead to low yields and clog the cage (a part of the press). Moist seed may also get moldy.

— *Clean:* Fine dust in the seed may clog the cage. Chaff left in the seed will absorb some of the oil and keep it from getting squeezed out of the cage. Sand in the seed will wear the press out. Stones badly damage the piston.

— *Warm:* Warm seed will yield the most oil for the least effort.

Seed should be dry before it is bagged and stored. Very damp seed will feel humid when you bury your hand in it, especially if the seed is warm. If you heat your seed in the sun under a sheet of clear plastic, you may see moisture collecting beneath the plastic if the seed is too wet. Seed that is slightly too damp may feel dry but will not press well. If it is too damp, but not yet moldy, it can be dried in the sun.

Spread the seed out thinly on the ground, on plastic, or on roofing tin. At the end of the day, pile the seed up to keep it from absorbing moisture in the cool night air, and spread it out again in the morning. If there is any chance of rain, or if the morning dew is heavy, you will need to bag all the seed in the evening and put it back out the following morning. After two or more sunny days, the husks will be dry. Now bag the seed and store it for a week. In that time, the moisture in the seed will be drawn into the dry husk, and the entire seed will become evenly dry.

Winnowing is a low-cost way to clean the seed. On a day when there is a steady breeze, spread a cloth on the ground. Drop the seeds from a container to the cloth. The chaff and dirt will blow away. The seeds will collect on the cloth. Any stones must be picked out by hand. If the sun is shining, the seed can be heated on a piece of roofing tin or some black plastic. Spread the seed thinly and evenly on the sheet. On a sunny day, it will be hot to the touch in 15 minutes. The seed should be at least 104°F, which is pleasantly warm to the touch.

The yield will improve as the seed gets hotter, all the way up to 150°F, uncomfortably hot to the touch. The seed will heat faster on a black surface than on a light-coloured surface. Some people also cover the seed with a sheet of clear plastic. This helps to hold heat in and make the seed hotter. If the sun is not shining, you can also heat the seeds on a stove. Put the seeds in a shallow pan over low heat for five minutes. You will need to stir the seeds often and be careful not to scorch them. Charred seeds will give the oil a bad taste.

You can use a double boiler to make sure that the seeds do not burn. You will need two pots, one large and one smaller. Set the larger pot on the heat, with enough water to keep it from boiling dry. Set the smaller pot inside the big pot. The bottom of the small pot does not need to touch the water below. Put the seed in the smaller pot, stir it occasionally, and the steam from the boiling water will heat the seed without burning it. The seed should be ready in 15 minutes. The boiler works best if the smaller pot is almost the same size as the big pot. Then there is not too much room for the steam to escape from between them. This method does not require as much attention and stirring, and you will not spoil any seed from accidental charring, but it uses more fuel.

The seed is now ready to be pressed. The pressing operation is described in detail in the manual, complete with diagrams. Dr. Rita Laker-Ojok, a former EWW staffer who worked in Africa, cautions that knowing how to use the press is important, especially regarding pressure adjustments. Correct adjustment allows the "cake," or meal, to come out; too much pressure may cause a jam. According to Dr. Laker-Ojok, the sunflower oil produced is very high quality. It can be further processed, or clarified, in one of three ways. The simplest is to let it settle for a week. The oil can also be mixed with a little water and salt and boiled for 10 minutes. This changes the taste slightly. Or the oil can be filtered through paper or cloth.

Seed cake

Seed cake is a valuable by-product of pressing. Sesame seed cake is valuable as a human food. Sunflower seed cake is not suitable for people, but it makes a good addition to chicken, pig, or cattle feed. Since sunflower seed cake has all the seed hulls in it, is very fibrous. The ram press does not get all the oil out of the cake; it is oilier than most feed additives. It is quite high in crude protein, but contains very few carbohydrates. It should be used as a feed additive, not a feed by itself.

The EWW manual describes feed rations for commercial broiler chickens, cows, and pigs that include sunflower seed cake. Proper storage of both seed and seed cake is extremely important. Seed must be protected from moisture, rodents, and insects. Very moist seed will rot. Even if your seed is not moist enough to rot, it may be moist enough to grow mold. This is a problem for two reasons. First, moldy seed cake does not taste good to animals. They may not be willing to eat moldy feed.

Worse, some kinds of mold make mycotoxins such as afiatoxin. These poisons can make people and animals sick. Some of the poisons from moldy seed will end up in the oil, but most remain in the seed cake. They can also get into the meat, eggs, and especially the milk of the animals that eat the cake. Mold spores (seeds) are present in all crops. Molds grow best in warm, humid weather. To prevent the growth of mold, dry the seeds shortly after harvest. Even dry seed can quickly get damp by being in contact with damp earth.

Once the seed is dried and bagged, it must be stored carefully to keep it from taking up moisture. The moisture content of the seed should be no higher than 10%. To test for moisture, weigh a sample of seed or cake, and then heat the sample in an oven at 300°F for one hour. Reweigh the sample. The weight lost in the oven is equal to the moisture content of the original sample, and the percentage can be calculated: divide the weight lost by the original weight and multiply by 100.

Other Types of Oil Presses

Several types of small-scale extractors are commercially available, both imported and manufactured in the U.S. Two that are commercially available in the U.S. are described here.

Komet Oil Expellers

Komet Vegetable Oil Expellers are manufactured by IBG Monforts in Germany, whose range of products covers small hand-operated as well as industrial machines. According to the product literature, Komet oil expellers feature a special cold pressing system with a single conveying screw to squeeze the oils from various oil-bearing seeds. The machines operate on a gentle mechanical press principle that does not involve mixing and tearing of the seeds.

Virtually all oilbearing seeds, nuts, and kernels can be pressed with the standard equipment without adjusting the screws or oil outlet holes. The vegetable oil produced needs no refining, bleaching, or deodorising, as long as the natural taste, smell, and colour are acceptable. Generally, any sediment in the oil will settle to the bottom of the collecting vessel after approximately 24 hours and form a hard cake. You can then pour the oil through a paper or textile filter to remove any remaining impurities.

Since the vegetable oil will not reach temperatures higher than approximately 100 to 120°F, all its vital components will be preserved. Therefore, it is excellently suited for natural nutrition. As long as the oil is stored in a dark, cool place, it will have a long shelf life. You must precondition the seeds before extraction. Big nuts, kernels, and copra (dried coconut meat) have to be crushed to the size of peas on the Komet Cutting Machine "System Crusher."

The Täby Press is a screw press manufactured in Sweden. Various models are available for coldpressing rapeseed, linseed, faxseed, sunflower seed, sesame seed, peanut, groundnuts, mustard seed, poppy seed, cotton seed, jojoba, etc. Bengt Jonsson is the designer, manufacturer, and seller of the oil press, and is also a farmer. The Type 20 screw press is distributed in the U.S. by Magic Mill International.

Value-added Processing

Although value-added processing can be a way to increase a farmer's share of the farm product dollar, adding value doesn't come free. At each step, the farmer is adding more work, buying additional equipment and supplies, and using more energy. Information needs increase. And depending upon the situation, liability insurance may become a must.

A key point to remember is that adding value to any foods by processing increases safety risks, due to the increased handling. Therefore, rules and regulations are established by each state to protect the public health. In addition, the FDA regulates products going out of state. Farmers considering processed foods as a value-added business should contact their state health department before proceeding.

OIL SEED PROCESSING PLANT

Oil Seed Processing has become a major industry through out the World Economy as Food Producers continue to develop more uses for these Edible Oils. Oil can be derived from such seeds as Corn, Soybean, Cotton, Canola, Peanut, Sunflower. There are approximately 272 Million MT (metric tons) of Oilseeds produced annually in the world and of this production over half the oilseeds grown are soybeans and 74 Million MT of Soybeans are produced per year in the United States.

Within the last few years, the emphasis of the Oilseed Oil Producer has changed from stand-alone independent operations toward the integrated manufacturing facility, producing a more complete range of value-added products from the raw seed to the dinner table. End products such as Salad Oil, Shortening, Mayonnaise, and Margarine are commonly produced from these process plants.

Modern Oil Seed Processing consists of a series of unit processes that take the seed through the following steps:

1) Drying
2) Cleaning, Cracking, Dehulling
3) Conditioning and Flaking
4) Solvent Extraction and Solvent Removal
5) Degumming
6) Alkali-Refining
7) Bleaching
8) Hydrogenation
9) Deodorising
10) Blending and Storage

Safety Procedures

Typical industry safety procedures are adhered to in most of these type plants. Care should be taken to wear safety glasses at all times and proper ear, head and foot protection is recommended. In the Extraction area of the plant where Hexane is present, care should be taken to eliminate any items that may cause a spark, which can result in a flash fire due to the highly flammable nature of this solvent.

Other areas of these plants may have steam present for product heating. In the Refinery care should be taken when moving around the plant as some of the floors can be slippery due to product leakage and caustic cleaning procedures. While in the packaging areas where food grade material is present, hairnets may also be required.

Overview of the Process

Dry end

Seeds are harvested and brought to the plant via rail or truck and transferred to the storage silos via mechanical and/or air assisted elevators. The seeds are then moved to the cleaning process via mechanical augers and conveyors where any field trash is removed. The seeds are dried to the desired moisture content of 12% and then tempered which causes the husk to pull away from the meat and prepares the product for cracking and dehulling. The seeds are then weighed into the rotary cracking equipment. The objective of cracking is to break the seeds into suitable pieces for dehulling and flaking.

The cracking is accomplished by passing the product through corrugated sawtooth opposing rolls running at varying speeds. The seeds now pass to either the vibratory screens, cyclone separators, or vacuum separator (Kice Separator) where they are subjected to up flows of dry air and the hull (husk) and meat (beans) of the seeds are separated. Cyclone Separators are utilised to remove the fine particles and trash from the hulls and collected in the bag house. The hulls are then sent to a secondary dehulling process, which separates the hulls into sizes for processing into meal and feed stock. Blowers and Vacuum units are used in this area to convey both materials to the next phase of the operation.

Conditioning

The beans or meat are then passed through a steam heating process. This is accomplished by either a vertical stack cooker or rotary horizontal cookers where the cracks are heated to approximately 72°C (161°F) and steam, or a water spray, is added to adjust moisture to approximately 11%, which makes the material plastic for flaking. On these types of cookers there are rotary steam joints which require sealing.

Flaking

Flaking mills consist of a pair of smooth-surfaced rolls of approximately 20 to 32 inches (508mm to 812.8mm) in diameter which operate in opposing directions with minimum clearance and driven at different speeds. Pressure between the rolls is generated by mechanical or hydraulic systems and flake thickness is controlled by adjustments of this pressure. Flake thickness is generally in the range of 0.005 to 0.020 inches (0.13mm to 0.5mm).

Expanding

The dry flake is then transferred to the Expanders (extruders) where water and steam is introduced into the product and mixed in the chambers of the extruder. The wet cake is then extruded from the expander at approximately 104°C (220°F), sized and cut via a die head to provide the optimum material for the extraction phase of the process. These extruders require shaft sealing on the drive end where the shafts tend to operate at slow speeds but with high pressures and solids content in the extruder chamber. The expander process is used when increased bed drainage is desired in the Extraction Process, mainly in deep bed or Rotocel extraction.

Solvent extraction

Solvent extraction of oilseeds is a diffusion operation in which the solvent (Hexane) selectively dissolves miscible components (oil) from other substances. The extractor unit provides the physical means of contact between the solvent and solids consisting of prepared oilseeds. Immersion and Percolation are the most common types of extraction though Percolation extraction is more common in modern process plants.

The Percolation method utilises a bed composed of oilseed flakes and counter current flow of the flakes and miscella (the mixture of oil and Hexane). Three types of extractors are commonly found in the industry:

1) Rotary or Deep Bed,
2) Horizontal Belt,
3) Continuous Loop.

A large percentage of the modern process plants utilise the Horisontal Continuous Loop though some of the older facilities still rely on the Rotary or Deep Bed technology. The Horizontal Continuous Loop units utilise a conveyor system which moves the miscella through different stages of solvent introduction and screening, and finally a product dump cell. The Rotary Deep Bed units consist of a series of concentrically arranged cells, which are filled with oil-bearing material. Each cell is filled consecutively and brought in contact with decreasingly concentrated miscella.

The freshly charged solids are brought into contact first with the most concentrated miscella and the fully extracted solids contact fresh solvent before discharging. Sealing of the rotating shaft on both top and bottom has historically been accomplished with mechanical packing, air energised boot or lip seals. Non metal contacting seals are preferred in this area due to the presence of Hexane in the process and the desire to minimise any sparking due to metal to metal contact or heat generation.

Solvent Removal

The spent Hexane solvent must be recovered and recycled with a minimum of loss. Solvent leaves the extractor as full miscella and as solvent in the flakes, and is then transported to desolventising. Solvent is removed from the Miscella by double-effect evaporation and steam stripping via a rising film evaporation process. In the first effect evaporator the mixture is 80% oil and the mixture is heated to approximately 38°C (100°F). In the second effect evaporator the mixture is heated with 182°C (360°F) steam to drive the Hexane out and leave a mixture of 98% oil at 110°C (230°F).

The miscella is then contacted directly by steam in the primary stripping operation and the Hexane is again driven off leaving 99.5% oil. In the secondary stripping process, steam is again injected into the miscella and the final 99.9% pure oil product is produced. Producers strive for less than 100ppm of Hexane in their pure oil products. Hexane recovery is essential and mandated by the EPA. The remaining Hexane present in the non-condensable gases is removed via a mineral oil stripping process.

Purchased mineral oil is mixed with the Hexane latent gases in the absorption column where the Hexane is absorbed by the cold mineral oil. The cold mineral oil latent with Hexane is then heated and pumped to the top of the stripping column. Steam from the stripper passes to a condenser for recovery of the solvent. Solvent free gases from the absorber are vented to the atmosphere through a steam ejector or fan. The recovered Hexane is recycled by the facility and mixed with fresh Hexane to be re-used in the extractor process.

Degumming process

Oils high in phosphorus, such as soybean, corn, and sunflower, may be degummed prior to refining. The degumming process is not a mandatory process due to the fact that the phosphatides (lecithin) can also be removed in the subsequent processes. The primary reason for degumming is to provide a crude-degummed oil suitable for storage or long transit or to produce lecithin to be sold as a by product for addition to livestock and poultry meal feed to increase the energy content.

Alkali-refining process

Alkali-Refining is the process designed to neutralise free fatty acids present in the oil. If degumming is not included in the facilities process, then caustic refining is the traditional first step for edible oil processing. This refining phase is important to remove the impurities in the oil, which have a tendency to turn dark when heated in the subsequent stages of the process. As the crude oil is delivered from the extraction plant, the oil is measured, cooled and injected with a measured amount of phosphoric or other acid. This acid has an affinity for the phosphatides (lecithin) and after bonding the acid and the phosphatides are removed from the process oil by a water wash.

The oil and acid is brought in contact with the caustic wash in a high shear mixing application and passed through controlled heat exchangers and centrifuges where the Raw Oil and Soap Stock (Acid and Caustic) are separated. The Soap Stock is then sold as a by product or sent to waste water treatment. The oil is transferred to the next phase of the process.

Bleaching process

Bleaching is the term given to describe the absorption cleansing process associated with edible oil refining. This process may include acid pretreatment, introduction of a retention aid and bleaching agent and the removal of the clay and absorbed materials. Other materials may also be introduced at this point, such as activated carbon for canola and other oils and filter-aids such as Diatomaceous earth, trisyl (small glass beads) and most recently rice hulls. The functions of the bleaching step is to remove or reduce the levels of pigments, oxidation products, phosphatides, soaps and trace metals in the oil.

The bleaching process improves the colour of the oil. These clays and bleaching agents may also be introduced via a slip stream line and inline style of mixer. Between the slurry tank and the bleaching vessel the oil is heated to prepare it for the bleaching process. The oil is then agitated in the bleaching vessel for several minutes and then passed through the bleaching filter to remove any solid materials. These bleaching filters may be either plate and frame or leaf filters. As these filters become blinded, they are taken off line and cleaned manually or by vibratory action. The cake is collected and mixed with the meal at the dry end of the process. The oil is then passed on to the hydrogenation and deodorising processes.

Hydrogenation

Hydrogenation is generally performed for one of two specific purposes. The first is to provide taste and smell stability and to enhance shelf life for unsaturated products. The second is to change the functional characteristics of the naturally occurring fats to those required for specific applications and functions. In the hydrogenation process the bleached oil is transferred to a reaction vessel where it is heated to 163°C (325°F) or higher to promote the reaction to commence. Water vapor is removed from the top of the vessel and catalyst (usually nickel) and hydrogen gas is added to the bleached oil and agitated per the individual processor recipe. When the processor has reached the desired results, the oil slurry is then passed through an oil-oil economiser to bring the hard oil to filtration temperature. Filter aids such as Diatomaceous earth are again added in this stage. The filter aid and hard oil slurry is then passed through the leaf press filter to remove the nickel catalyst and filter aid. The captured catalyst is recycled and the hard oil is sent through a finishing filter and on to storage or to the deodorisation stage of the process.

Deodorisation

Deodorisation in the processing and refining of edible oil products is used to prepare the oil for use as an ingredient in margarine, shortening, salad oil, cooking oil, hard butters for the confectionery industry and many other products in the food industry.Deodorisation is a steam stripping process wherein a good-quality steam, generated from deaerated and properly treated feedwater, is injected into the oil under low absolute pressure and sufficiently high temperature; 260°C (500°F), to vaporise the free fatty acid and odoriferous compounds and carry these volatiles away from the feed stock oil.

REFERENCES

Gulati, A., and T. Kelley, *Trade Liberalization and Indian Agriculture*, Oxford University Press, 1999.

Gulati, A., A. Sharma, and DS Kohli, "Self-Sufficiency and Allocative Efficiency: Case of Edible Oils," *Economic and Political Weekly*, 31(13): A14-A24, 1996.

Murthy, K.N., *Inter regional variations in consumption patterns in semi-arid tropical India: A demand system approach*, Indian Statistical Institute, Delhi Centre, Discussion paper 01-06, 2001.

Herz, Jonathan, "Using and Maintaining the Ram Press," *Enterprise Works Worldwide*, Washington, DC, 42 p, 1997.

3

Production and Marketing of Sunflower Oilseeds

Sunflower is one of the most important field crop cultivated in Africa and Asia. Over the past ten years, the area planted under sunflower has fluctuated drastically. An important relationship exists between the area planted under maize and the area planted under sunflowers due to the nature of their substitutability. Sunflowers are well adapted to hot and dry climate and can be produced economically even when there is not enough moisture to produce most of the other summer crops.

Approximately 95% of all sunflower seed produced is destined for the processing industry for the production of sunflower oil. Sunflower oilcake - also referred to as meal -, a by-product of the oil extraction process, is sold to feed manufacturers domestically. It is generally regarded as a low-value product that does not compare well to soybean oilcake in terms of nutritional value. Therefore, the dilemma of the sunflower market is that not enough sunflower seed is produced locally for the oil industry, yet once the seed has been crushed the by-product is regarded as a low-valued product.

GROWTH AND DEVELOP OF SOYBEANS

It is important to understand how soybeans grow and develop. Since their vegetative and reproductive growth stages occur for several weeks, many environmental conditions can affect final yield. Too much or too little moisture at specific stages of growth can affect performance. Even the type of soybeans, i.e. determinate or indeterminate, influences when a plant starts the flowering or pod-filling stage.

Stages of Growth

A descriptive system has been worked out for soybean growth and development. Table 1 describes the stages. Root growth begins at germination. Primary and lateral roots grow strong until R5. After R5, the shallower roots degenerate, but the deeper roots and laterals grow until R6.5. Nodulation occurs as early as VI and continues through the V stages. Nitrogen is fixed by the rhizobium and is fed to the plant. Phosphorus and potassium are taken up by the roots - slowly at first, then rapidly. Rapid nutrient uptake occurs before a period of rapid growth. Most nutrients are taken up by the time the plant reaches R6 stage.

Stage R4 marks the beginning of the most crucial period of plant development in terms of yield determination. Stress (moisture, high temperature, nutrient deficiencies, lodging, or hail) occurring from R4 to R6.5 will reduce yields more than the same stress at any other period of development. Stages R4.5 through R5.5 are especially critical. Young pods are more prone to abort under stress than older pods and seeds. Stresses at this stage can greatly affect final performance.

Selection of Soybean Varieties

The choice of variety is a key factor in profitable soybean production. Many varietal characteristics, such as maturity, lodging, and disease resistance, must be considered when selecting varieties to complement a production area.

Table 1. Description of soybean growth stages

Stage No.	*State Title*	*Description*
Vegetative Stages:		
VE	Emergence	Cotyledons above soil surface.
VC	Cotyledon	Unifoliate leaves unrolled so leaf edges are not touching.
VI	1st node	Fully developed leaves at unifoliate nodes.
V2	2nd node	Fully developed trifoliate leaf at node above unifoliate node.
V3	3rd node	Three nodes on main stem with fully developed leaves beginning with the unifoliate node.
V(N)	nth node	n = number of nodes on main stem with fully developed leaves beginning with the unifoliate node.
Reproductive Stages:		
R1	Beginning flowering	Open flower at any node on main stem. Indeterminate plants start at bottom and flower upward. Determinate plants start at one of the top 4 nodes and flower downward.
R2	Full bloom	Open flowers on one of the two uppermost nodes on main stem.
R3	Beginning pod	Pod 3/16 inch long at one of the four uppermost nodes on main stem.
R4	Full pod	Pod (3/4 inch) long at one of the four uppermost nodes on main stem.
R5	Beginning seed	Seed (1/8 inch) long in one of the four uppermost nodes on main stem.
R6	Full seed	Pod containing a green seed that fills pod cavity one one of the four uppermost nodes.
R7	Begin maturity	One normal pod on main stem has reached mature pod color.
R8	Full maturity	95% of pods have reached mature pod color. Approximate 5 to 10 days ahead of harvest.

Maturity

Maturity must be closely matched to the production environment and the cropping system. Maturity strongly influences how plant development complements or is affected by climatic conditions.

Soybean plants are sensitive to day length or photoperiod. The plants' response to day length controls the timing of the transition from vegetative to reproductive or floral development and the rate of physiological development. Varieties differ in their responses to day length. Some varieties flower under relatively short days while others flower under longer days. Varieties have been classified for photoperiod response based upon the ability of the variety to effectively utilize the length of the growing season in a region.

During the summer, day length increases from thesouthern to northern United States. This change in photoperiod results in regions of adaptation in the form of horizontal bands running east to west across the United States from the northern states to the Gulf Coast area. Each band extends 100 to 150 miles north to south. Varieties adapted for full-season production to a particular region are assigned a maturity group number. Varieties in group 00 are adapted to the northernmost regions of North Dakota and Minnesota, and those in group VIII to the southernmost region of Florida and the Gulf Coast states. As the maturity group number increases, the lengths of the vegetative and reproductive stages of development are extended. A spread of about 10 days in maturity exists for varieties classified in the same maturity group.

Varieties are often moved north or south of their primary area of full-season adaptation to complement a production area. Because of the change in day length, when a variety is moved to the north, flowering will be delayed, and when it is moved to the south, flowering will be hastened.

While planting too late a variety will increase the likelihood of frost damage, switching to a substantially earlier maturing variety should be resisted for two reasons. First, early maturing varieties planted late in the season will usually have limited vegetative development, short stature, and low yield potential. Second, the number of days to flowering, pod development, and maturity will be less in later plantings than earlier plantings of the same variety. At the time of planting double-cropped soybeans, the day length has begun or will soon begin to shorten. This reduction in the day length, or the increase in the night period, causes plant development to speed up. Consequently, planting the same variety at the end of May versus the end of June does not delay maturity by one month. As a general rule, for every 3 days delay in planting, maturity is delayed by only 1 day. Since soybean development is hastened in later plantings, the highest yields in a double-cropped system are often achieved by the same variety or one slightly shorter in maturity than what is used in full-season production.

With environmental conditions fluctuating from year to year, planting varieties that vary in maturity should be considered by growers. Utilizing a range of maturities can reduce the risks associated with weather-related yield reductions and facilitates timely harvest and efficient machinery use.

Standability

A variety must be able to remain erect throughout the growing season. Lodging during the vegetative or reproductive growth will disrupt the light penetration into the plant canopy and may reduce seed yield. Lodging late in the season may reduce harvest efficiency and increase harvest losses. While varieties differ in their ability to resist lodging, environmental conditions greatly influence the tendency to lodge. Factors such as irrigation and high fertility tend to promote vegetative development and increase lodging. Increasing plant population causes the stems to become taller, more slender, and more prone to lodging. If lodging has been a problem in the past, the grower should consider selecting a more lodging-resistant variety in addition to re-evaluating the seeding rate at planting.

Stem Termination Type

Indeterminate varieties continue to grow vegetatively several weeks after flowering begins. Plant height can more than double after floral initiation. In contrast, determinate varieties complete most of their vegetative growth before flowering. Plant height can increase slightly in determinates after the onset of flowering, but generally 70 to 80 percent of the mature plant height is achieved prior to flowering.

At maturity the indeterminate plant tends to have a relatively even distribution of pods on the stem with a lower frequency toward the tip of the stem. The determinates tend to have a dense cluster of pods on the terminal raceme. Comparing varieties of similar maturity, indeterminates are usually taller than determinates. While most of the varieties in groups 00 through IV possess an indeterminate growth habit and varieties in groups V to VIII possess the determinate growth habit, exceptions to this rule do occur. A few group IV and earlier varieties have been released that are determinates and a few indeterminates of group V or later maturity have been released.

The determinate growth habit in an early-maturity background results in short, compact plants that tend to have excellent lodging resistance. These early determinate (sometimes referred to as semi-dwarf varieties) varieties are recommended for production systems involving narrow rows, high seeding rates, early plantings, good fertility, and a yield potential in excess of 50 bushels per acre. While these early determinates can be productive, unless lodging is a problem, these varieties are not expected to be higher yielding than the best indeterminate varieties.

The determinate growth habit was originally incorporated into varieties to reduce plant height and lodging in environments with long growing seasons. The indeterminate growth habit, which has been incorporated into a few group V and later maturity backgrounds, tends to result in a tall, bushy variety with a long flowering period. These late indeterminates have been recommended for production in stress environments or double-cropped situations where vegetative production may be limiting.

Disease Resistance

A wide range of environmental conditions prevail during the growing season. These conditions influence the occurrence and severity of diseases attacking soybeans. Although numerous pathogens can affect soybean production, soybean cyst nematode (SCN), charcoal rot, and Phytophthora root rot are three of the more damaging soybean diseases. No single soybean variety can provide complete protection against all three of these diseases. However, a number of control measures involving variety selection can be used to keep losses due to these diseases to a minimum.

SCN population density and race structure affect the ability of the nematode to cause damage. Race is determined by the number of cysts that are produced on four soybean differential genotypes: Peking, Pickett, PI88788, and PI90763 (Table 2).

Resistant varieties are recommended when soybeans are grown in SCN-infested conditions with population density levels greater than the economic threshold.

Rotations that include nonhosts and susceptible varieties also are recommended to control population density. Proper use of resistant varieties requires that producers know the race of the nematode present. As mentioned earlier, different sources of resistance have been used to develop new varieties.

Table 2. Partial representation of the SCN race scheme

	Soybean Differential			
Race	Pickett	Peking	PI88788	PI90763
1	-	-	+	-
3	-	-	-	-
4	+	+	+	+
6	+	-	-	-
14	+	+	-	+

+, ≥ 10% of reproduction of the susceptible standard, Lee
-, < 10% of reproduction of the susceptible standard, Lee

Most resistant varieties will not prevent the buildup of all races of SCN, so knowing the race and monitoring the population density and race over time is important in preventing losses to the pathogen. Once the population density and race are known, a variety can be selected that will best prevent the buildup of the population.

Charcoal rot is an important soybean disease that is associated with drought stress and high temperatures. The most recent estimates quantify the U.S. production loss attributed to charcoal rot at more than 30 million bushels per year. Only losses from soybean cyst nematode and Phytophthora root rot exceed the estimates for charcoal rot nationwide. Charcoal rot is the only major soybean disease without good genetic resistance. Advances in recent research on charcoal rot disease have demonstrated that varieties differ in their reaction to infections by the pathogen, but higher levels of resistance are needed to effectively combat the disease. While no completely resistant varieties have yet been identified, the differences noted among varieties provide hope that high levels of resistance can be identified in plant introductions or other varieties.

Currently, the best recommendation for minimizing losses to charcoal rot is to select planting dates and variety maturity combinations that minimize drought and heat stress experienced by the plants and produce the highest yield. Final yields are a good indication of the levels of drought and heat experienced by the crop during critical periods of development.

Phytophthora root rot does not represent a widespread problem, but isolated areas of infestation do occur occasionally. Varieties are available that provide a high level of resistance to Phytophthora. Like SCN, Phytophthora is an organism that is present in fields in a number of different forms or races. More than 25 races of Phytophthora have been described. When losses to the pest are known to be serious, a variety with specific resistance to multiple races of the pathogen should be selected. Each resistance gene will control some, but not all, races.

Blends

Along with a large choice of varieties, producers also may select from a number of blends (mixtures of two or more varieties) being marketed. The process used to select an appropriate blend is similar to that used in varietal selection. However, unlike varieties that maintain their genetic identity season after season with proper handling, the frequency of the individual components of a blend may change after only one season in production. If this occurs, the basic characteristics of the blend also may change, affecting performance. To ensure a blend contains the correct proportion of components each season, the components must be mixed and the seed purchased each year.

Shattering

Environmental conditions prevalent during the harvest season often test a variety's ability to resist shattering. As a regular component of Soybean Variety Performance Tests, entries are evaluated for shattering resistance. These evaluations should indicate the relative shattering response of a variety when left in the field about 2 weeks after maturity or about 1 week later than the optimum harvest time. To minimize shattering in blends, select a brand in which the earliest and latest maturing components do not differ by more than a few days.

Iron Chlorosis

Iron chlorosis is a common problem on highly calcareous soils. Although no soybean varieties are available with complete resistance to iron chlorosis, moderate levels of tolerance do exist that allow for improved soybean production in all but the most severe problem areas. Often the developer of a variety will know how a particular variety responds to conditions favoring the development of iron deficiency chlorosis. Every few years researchers with the KAES conduct variety evaluations for chlorosis tolerance.

Yield Potential

Harvestable yield is obviously an important characteristic to consider when selecting a soybean variety. Although improvements in yield do not occur at a rapid pace, development of new, disease-resistant varieties and integration of those varieties into cropping systems account for approximately half of the improvements in soybean yields. This genetic contribution represents an increase in yield of 0.1 to 0.2 bushels per acre per year. Each year many newly released varieties will out-yield previous releases, particularly older varieties, by 10 to 20 percent in various production situations.

To make the best possible decision regarding yield potential, consult results from public trials, county test plots, farmer evaluations, and private tests for information on variety performance. Since the performance of any variety will vary from year to year and from location to location depending on factors such as weather, management practices, and variety adaptation, carefully consider the performance of a variety across several locations for 2 or more years. An average across environments will provide a better estimate of genetic potential and stability than a single or only a few test sites.

Selecting a variety or a number of varieties for production is one of the many decisions the soybean producer must make throughout the year. The substantial differences in yield potential among varieties under various conditions make this decision one of the most important. Final selection should be based on a thorough understanding of the production conditions and a careful assessment of varietal characteristics that best complement those conditions.

SEEDBED PREPARATION

Soybeans need a soil that is warm, moist, wellsupplied with air, and provides good contact between the seed and soil for rapid germination. An ideal seedbed should accomplish the following:

— control weeds,

— conserve moisture,

— preserve or improve tilth,

— control wind and water erosion, and

— be suitable for available planting and cultivating tools.

Seed Quality

Soybean seeds are extremely fragile and subject to damage by handling, augering, and transporting. Seed moisture influences the extent of damage; drier seed is more easily damaged than seed with higher moisture content. Splits and cracks in seed can be seen easily, but considerable internal injury can go undetected. The only way to determine seed viability is to have a germination test run. The germination rate should be 85 percent or greater.

Row Width

Most producers use a 30-inch row width since it is well suited for other row crops, allows cultivation for weed control, and out-yields wider row spacings. Under irrigation, the irrigation method dictates the row width.

Row widths narrower than 30 inches, although not a new concept, have several advantages. They provide early canopy cover, improved light interception, better plant or stand survival, ease of harvest, and increased seed yields. The narrow row soybean yield advantage has been quite variable. Stress reduces or eliminates any gains from narrow row spacing, but under irrigation, the increased yield advantage is greater and more consistent. Problems with poor seedling emergence or stand establishment often are associated with narrow row spacings since there are fewer seeds per foot of row to push through a crusted soil. Uneven seed metering by the drill or poor seed-soil contact also can cause stand problems. Press wheels are needed to improve the seedsoil contact and aid germination and seedling emergence.

Early-maturing varieties tend to respond better to narrow row widths than later-maturing varieties, although they may not be as productive. Varieties that perform well in wide rows usually perform well in narrow row spacings.

Current research shows narrow-row (7 to 8 inches) beans are equal to wider rows (30 inches) when weeds are controlled and plant population per acre is not increased. A 10-percent yield increase is noted for narrow rows over

30-inch rows when planted after June 5. It is very important to keep plant population the same on a per-acre basis. Note that 105,000 plants per acre in 30-inch rows is equal to six plants per foot of row. In 8-inch drill rows, that same population per acre equals 1.6 plants per foot of row. Do not overplant when drilling soybeans. Because of poor seed-soil contact with some drills, seeding rates may need to be increased by 15 percent over that with the row planter. In high-yielding conditions (greater than 40 bushels per acre), seeding rates may again need to be increased.

Plant Population

Since soybean seed size varies among varieties and within the same variety grown in different environments, it is essential to consider planting rates in terms of seeds per foot of row or seeds per square foot. It is not uncommon to see sizes ranging from 12.6 grams per 100 seeds to 18.9 grams per 100 seeds (2,400 to 3,600 seeds per pound). If one variety has 2,400 seeds per pound and another 3,600 seeds per pound, and a producer planted both at a rate of 60 pounds per acre, the large-seeded variety would have a seeded population of 144,000 seeds per acre while the small-seeded variety would have 216,000 seeds planted per acre. If 60 pounds of both varieties were planted at nine seeds per foot of row in 30-inch rows, the largeseeded variety would plant 0.92 acres while the smallseeded variety would plant 1.37 acres.

To minimize seed costs, producers need to know precisely the number of seeds planted per foot of row. There are two simple equations to calculate the final seeded or plant population and the number of seeds or plants per row-foot to obtain a desired plant population. These equations, shown later, are helpful in determining the amount of seed to purchase and in calibrating planting equipment.

The optimum planting rate may be adjusted for different growing conditions such as upland versus bottomland, irrigated locations, and planting dates. Interplant competition increases with increased plant populations resulting in taller plants with smaller stem diameters, which accentuates lodging problems and fewer branches, pods, and seeds per plant. Low plant populations or thin stands cause low branching and pods produced lower on the stem, which increases harvest losses.

Since early-maturing varieties generally do not produce a dense canopy, the planting rate should be adjusted upward. Other situations such as upland versus bottomland sites or late plantings may require the planting rate be adjusted either up or down. Drier soil conditions usually require a lower planting rate while late plantings require increased rates to compensate for the lack of canopy growth.

Planting Depth

The optimum planting depth for soybean seeds is 1 to 1.5 inches, no more than 2 inches in sandy soils. With early plantings, slightly shallower seed placement will speed emergence, and with late plantings in dry soil, slightly deeper placement may be necessary to put the seed in contact with moisture. The use of press wheels will help firm the soil resulting in good seed-soil contact for rapid germination and uniform emergence. Certain herbicides necessitate deeper placement to reduce seedling injury.

Soybeans can emerge from plantings deeper than 2 inches, but seedling emergence is slowed. The seedlings are subjected to more disease organisms, and poor stands frequently occur on heavy soil from surface crusting. When soil crusting occurs, the hypocotyl of the seedling is not able to break through the crust. If the crust is left undisturbed, the hypocotyl arch or crook will break or the seedling will grow laterally, never emerging. A rotary hoe is useful for breaking a crusted soil and allowing the seedlings to emerge, but it will reduce the plant population slightly.

Double Cropping

Double cropping is the practice of planting a second crop after the first crop is harvested. If double cropping is to be successful, soybeans must be planted immediately after small-grain harvest.

Available growing season and adequate moisture are often limiting factors. Substantial yield reductions occur when planting is delayed from mid-June to mid-July. Since growing time and moisture are limiting, many producers burn the small-grain stubble and plant using a no-till planter or till lightly with a disk or field cultivator and then plant.

More producers are using no-till planters and planting directly into the small-grain stubble, which saves both time and moisture. Major tillage operations using a moldboard plow or offset disk delay planting and cause a loss of valuable soil moisture. Some soils may require tillage to eliminate hard pans or compaction problems prior to planting.

Double-cropped soybeans may experience poor early growth due to excessive small-grain straw. This poor growth and leaf chlorosis can be corrected or avoided with a light application of nitrogen.

Conservation Tillage

A number of tillage and planting systems can be used in producing soybeans. These systems may include primary or secondary tillage, both, or no preplant tillage. Tillage and planting systems that provide for profitable crop production with minimal soil erosion are often referred to as conservation tillage systems, an umbrella term that includes reduced-till, mulch-till, ecofallow, strip-till, ridgetill, zero-till, and no-till methods. The emphasis in conservation tillage is erosion protection; however, savings in water, fuel, labor, and equipment may be additional benefits. Conservation tillage is an integral part of many conservation plans for highly erodible land (HEL) as a result of the conservation compliance provisions of the 1985, 1990, and 1995 farm bills.

The chance for erosion is greater on soils planted to soybeans than most other crops. Soybeans leave the soil surface loose and mellow, which can contribute to greater erosion losses. Soybeans produce less residue than most other crops, and the residue is fragile and easily broken down. Soybeans typically produce 45 pounds of residue per bushel of grain, compared to 60 pounds of residue for corn and grain sorghum and 100 pounds of residue per bushel of wheat. Very little soybean residue is standing after harvest. Standing residue is more effective than flat residue in reducing wind erosion losses. For these reasons, conservation tillage or other erosion-control practices are very important before, during, and after soybean production. Conservation tillage systems protect the soil surface from the erosive effects of wind, rain, and flowing water either by covering the soil surface with crop residue or growing plants or by increasing the surface roughness or soil permeability.

A common goal of conservation tillage systems is to reduce soil erosion losses below the soil loss tolerance or "T" value. Soil loss tolerance is an estimate of the maximum annual rate of soil erosion that can occur without affecting crop productivity over a sustained period.

The amount of residue necessary for protection against wind and water erosion depends on several factors such as climatic conditions and patterns, soil erodibility, surface roughness, field length, length and steepness of slope, cropping practices, and other conservation practices. A rule of thumb suggests leaving a 30-percent residue cover on the soil surface after planting where water erosion is the primary concern. Where wind erosion is a problem, 30 percent residue or its equivalent is required on the soil surface during the critical wind erosion period. The actual level of residue required to minimize soil loss on a specific field may vary.

Local NRCS personnel can provide assistance in determining residue needs. In situations where conservation tillage alone may not adequately protect the soil from erosion losses, it can be integrated with other conservation practices such as terracing, contouring, strip cropping, or windbreaks to provide the necessary erosion protection.

Several experiment fields and stations have studied the effects of various conservation tillage practices on soybean yields. In most cases, the form of tillage used did not influence yields significantly. However, in some years no-till soybean yields were lower than yields from reduced or conventional tillage. The reduction in no-till soybean yields occurred when soybeans were grown continuously or on poorly drained soils. Ridgetill is usually a good alternative on poorly drained soils. Switching to a crop rotation system and avoiding no-till on poorly drained soils as well as obtaining adequate plant stands, weed control, and uniform residue distribution from the previous crop are all important in conservation tillage soybean production.

FERTILIZATION

Fertile soils that produce high yields of other crops also will produce high yields of soybean. Total nutrient uptake by soybean depends on yield obtained, which will vary with season, variety, soil, and cultural practices. Soybeans take up relatively small amounts of nutrients early in the season, but as they grow and develop, the daily rate of nutrient uptake increases. Soybeans need an adequate supply of nutrients at each developmental stage for optimum growth.

High-yielding soybeans remove substantial nutrients from the soil, and this should be taken into account in an overall nutrient management plan. Soybeans contain a higher amount of potassium in the grain than does wheat, corn, or grain sorghum and thus remove more potassium.

Lime

The nodule bacteria (Bradyrhizobium japonicum) associated with soybean roots function best in soils of near-neutral pH. A near-neutral pH also favors availability of most plant nutrients. The need for lime can be accurately determined by a soil test. Lime recommendations are typically made for pounds of effective calcium carbonate (ECC) to bring the top 6 or 7 inches of soil to pH 6.8. For maximum effectiveness, the lime must be thoroughly incorporated into the soil. Lime is relatively insoluble and may take several months to complete the neutralization reaction. Lime application rates should be adjusted for tillage methods (incorporation depth). If shallow incorporation is done, then the recommended lime rate for the top 6 or 7 inches of soil should be adjusted.

Nitrogen

Under normal conditions, soybeans will need no nitrogen fertilizer as the nodule bacteria will fix sufficient nitrogen for optimum growth. If soybeans have been grown recently on the land and were well nodulated, reinoculation probably is not necessary. However, if there is any question about the abundance of nodule-forming bacteria in the soil, or if the land has no previous history of soybean production, inoculation of the soybeans is recommended. Handling, storage, and use should be in accord with the package label to assure viable bacteria in the inoculant.

Until nodulation occurs, the soybean plant depends on soil nitrogen for growth. Soybeans planted where a large amount of wheat straw has been freshly incorporated into the soil may respond to a starter application of 10 to 20 pounds of nitrogen. The wheat straw causes a temporary tie-up of the soil nitrogen by microorganisms decomposing the straw. On newly leveled land being planted to soybeans for the first time, an application of 30 to 40 pounds of nitrogen per acre may be necessary because of the extremely low available nitrogen in the soil. Recent research has shown that under high yield conditions (greater than 60 bushels per acre) an application of 20 to 40 pounds of nitrogen at the R-3 stage has resulted in a fairly consistent 5- to 10-percent yield increase.

Phosphorus

Phosphorus applications should be based on a soil test. Consistent responses to direct phosphorus fertilization generally have been restricted to soils testing very low or low in available phosphorus. With medium testing soils, responses have been erratic and normally quite small.

Limited work with placement has revealed little difference between planting-time banded starter and broadcast application methods. Phosphorus moves little in the soil, making it necessary to incorporate broadcast applications. If a starter fertilizer is used, the material should be placed to the side and below the seed. Soybean seeds are very easily injured by fertilizer, therefore, no direct seed contact with fertilizer is advised. No difference exists between phosphorus fertilizer sources.

Potassium

Soybean seeds are relatively high in potassium in comparison to corn, wheat, and grain sorghum. This means removal of potassium by soybeans is greater than for those crops on a per-bushel basis when the grain only is removed.

As with phosphorus, a soil test is the best index of potassium needs. Soils testing very low or low should be fertilized with potassium, either as a banded starter at planting or broadcast and incorporated. Potassium should not be placed in contact with the soybean seed because of possible salt injury. Yield increases from potassium are comparable to those with phosphorus under very low and low soil test levels.

Zinc

Soybeans have been found to respond to zinc application to soils testing low in available zinc. Zinc deficiency is most likely to occur under high-yield conditions, in cut areas of fields leveled for irrigation, or in eroded areas that are low in organic matter. The need for zinc should be determined by soil tests. Zinc fertilizer can be either banded at planting or broadcast preplant with little difference in response when applied at an adequate rate. Both organic and inorganic zinc sources (chelates and nonchelates) can be used. The chelates are considered three to five times more effective than the inorganic sources. Manure applications also are effective in eliminating zinc-deficiency problems.

Iron

Iron deficiency may be encountered on calcareous soil. Soil applications of iron materials presently available are not economically effective. For this reason, foliar sprays or manure applications are recommended to correct an iron deficiency. Again, soil tests can be used to predict the need for iron.

Other Nutrients

In general, soybeans are unlikely to respond to other nutrients. Calcium and magnesium are relatively abundant in most soils. Liming of acid soil supplies sufficient calcium so a deficiency of this element would not be expected.

Molybdenum deficiencies are more likely in strongly acidic soils than in neutral or only slightly acidic soils.

Liming acidic soils tends to make native molybdenum more available to plants. Thus, an adequate liming program should ensure against molybdenum shortages in most soils.

Research with sulphur, boron, copper, and manganese has not shown consistent responses. These elements should not be a problem for optimum soybean yields in most soils where soybeans are now being produced. If a deficiency of one or more of these nutrients is suspected, application should be made to a small test area to see if a response can be obtained. Producers should remember that several of these micronutrients are toxic in relatively small amounts beyond optimum levels. It is much harder to remove the excess than to add the nutrient.

WEED MANAGEMENT

Weeds growing with soybeans compete with the crop for light, moisture, and nutrients. Uncontrolled weeds reduce soybean yields and interfere with harvest. An effective weed-management program requires knowledge of the potential weed problems and implementing a timely weed-control program. Soybeans are very competitive with weeds once the soybeans develop a canopy, but early emerging weeds can cause serious problems. Thus, early-season weed control is the key to providing the soybeans with a competitive advantage and minimizing the effect of weeds. The most effective weed-control programs involve a variety of control practices in an integrated weed-management system, including crop rotation, cultivation, sound agronomic practices, and judicious use of herbicides.

Weed Problems

Summer annual grass and broadleaf weeds like crabgrass, foxtail, pigweed, velvetleaf, and cocklebur are common weed problems in soybeans. These weeds germinate in the spring and summer and produce seed before they die in the fall. Germination of summer annual weeds depends on soil moisture and temperatures. A welltimed cultivation or herbicide treatment can greatly reduce densities of summer annual weeds. Winter annual weeds like henbit, mustards, downy brome, horseweed (marestail), and mustards generally are not a problem in soybeans, except in no-till production. Winter annual weeds germinate in the fall and produce seed the next spring or summer. They can be controlled with tillage or burndown herbicide treatments prior to planting soybeans.

Perennial weeds that regrow each year from an established rhizome or root system are very competitive and difficult to control. Perennial broadleaf weeds such as field bindweed, common milkweed, and hemp dogbane are more difficult to manage in soybeans than perennial grass weeds like johnsongrass. Selective herbicides are available to help control johnsongrass, but are not readily available for control of perennial broadleaf weeds. Perennial broadleaf weeds are best managed by crop rotation or treatment between crops.

Preventive Weed Control

Preventive weed control measures are economical and prudent. Introduction of new weed species into fields can dramatically increase weed control costs. Weeds can be inadvertently introduced into clean fields with contaminated crop seed, tillage and harvest equipment, manure applications, animal movement, and flood waters. Some of these factors are controllable, while others are not. Planting weed-contaminated crop seed is one of the most common ways to spread serious weed problems to new fields. This is especially true for certain weeds that are difficult to separate from soybeans, like cocklebur, morningglory, and black nightshade. Soybean seed should be cleaned and inspected for weed seed before it is planted.

Clean equipment and combines when moving between infested fields. Root or rhizome segments of perennial weeds like johnsongrass or field bindweed can easily be transported within fields or between fields on field cultivators and other tillage implements. Combines are excellent at collecting and dispersing weed seed to other fields. If possible, harvest the clean fields or parts of fields before harvesting weedy areas, and clean the combine between fields.

Hand removal of scattered, newly introduced weeds may be a wise investment if done in time to prevent a serious weed problem from becoming established in a field.

Cultural Practices

Cultural practices can have a tremendous influence on the type and severity of weed problems. Crop rotation, tillage system, livestock wintering, and other fieldmanagement practices affect weed populations and competition in soybeans. Production practices that encourage quick soybean emergence and canopy development can give the crop a competitive advantage over many weeds. Proper seed placement, fertility management, planting date, and seeding rates can help establish a healthy, competitive soybean crop.

Planting configuration and seeding rate influence soybean canopy development. Solid-seeded soybeans tend to form a canopy more quickly and are more competitive with late emerging weeds than soybeans planted in wider rows. However, drilled soybeans cannot be cultivated, so earlyseason weed control becomes more critical. Soybeans planted in wider rows can be cultivated, which may reduce the need for herbicides.

Crop rotation is one of the most effective ways to manage certain weed problems specific to a crop. A serious infestation of velvetleaf, cocklebur, or devilsclaw in soybeans may be best managed by rotation to a different crop such as wheat, alfalfa, or corn, where the weed can be better controlled with tillage or alternative herbicides to lower the weed populations. Planned crop rotations can reduce weed, insect, and disease problems while improving soil tilth.

Tillage

Cultivation can be a cost-effective method of controlling weeds before and after planting soybeans. However, intensive tillage systems also increase the risk of soil erosion. Tillage implements that control weeds but maintain surface residues help minimize soil erosion. Cultivation is more effective for weed control with dry soils and sunny conditions than with moist soils and overcast skies.

Use of a rotary hoe after planting can control shallow germinating weed seedlings and alleviate soil crusting problems. Rotary hoes are most effective on small-seeded weeds like crabgrass, foxtail, and pigweed that germinate near the soil surface and are not yet emerged. Largeseeded weeds emerging from deeper in the soil, like cocklebur, velvetleaf, and shattercane, are much less susceptible to a rotary hoe. Some soybean stand may be lost during a rotary hoe operation. Soybeans are most susceptible to rotary hoe injury during the crook stage of emergence. The effectiveness of between-row cultivation depends greatly on soil conditions, weed and crop size, type of cultivator, and cultivator adjustments. Cultivation rarely provides complete weed control in the row. Cultivation and herbicides can be used together to reduce herbicide costs.

Herbicides

Herbicides, if used properly, are a safe and effective method to control certain weeds in soybeans. However, herbicides will not solve all weed problems and should be used only as needed in an integrated weed management program. Many herbicides are registered for use on soybeans to control different weeds. Important factors to consider when choosing a herbicide include:

1) weeds present,
2) stage of crop and weed growth,
3) herbicide persistence and crop rotation restrictions,
4) environmental considerations, and
5) herbicide costs.

Knowledge of the weed problems in a field and proper weed identification are essential when making herbicide decisions. Most herbicides selectively control certain weeds when applied as directed on the herbicide label. Weeds not listed on the label probably will not be controlled.

Herbicides are available for early preplant, preplant incorporated, preemergence, or postemergence application to soybeans. Products should only be applied at the recommended application timings to achieve the desired results.

Early preplant treatments. Some herbicides can be applied before soybeans are planted in the spring to control emerged weeds or for residual control of later emerging weeds. Emerged weeds must be controlled at planting for soybeans to be successful. If weeds are emerged at treatment time in no-till planted soybeans, a "burndown" herbicide that has foliar activity should be included in the treatment. Many residual herbicides applied early preplant need to be applied before weeds germinate and require precipitation or incorporation for herbicide activation. Early preplant herbicides need enough residual activity to control weeds before and after planting, or else need sequential treatments.

Preplant incorporated herbicides. Certain herbicides can be applied and incorporated into the soil before planting to control susceptible weeds. Preplant incorporated herbicides need to be incorporated thoroughly to the proper depth in the soil as directed on the label. Incorporation is more uniform with a dry mellow soil than with a damp or cloddy soil. Avoid furrowing too deep at planting time and moving too much treated soil out of the planted row, or weed escapes will occur in the row.

Preemergence herbicides. Some residual herbicides can be applied to the soil surface after the crop is planted, but prior to soybean and weed emergence. Rainfall or irrigation of about 0.5 inch is required after application to move the herbicide into the soil where it can be absorbed by the germinating weeds. Too little or too much rainfall following herbicide application can result in erratic weed control with preemergence treatments. A rotary hoe or rolling cultivator can be used to control small weeds and help incorporate the herbicide if insufficient rain is received to activate the herbicide.Preemergence herbicid es also can be applied in a band over the row at planting time. Band applications are generally 12 to 14 inches wide with a planned cultivation after crop emergence to control weeds between the rows. Banding reduces total herbicide use and cost.

Postemergence herbicides. Several postemergence herbicides are available to control emerged grass and broadleaf weeds in growing soybeans. Application stage, environmental conditions, and adjuvants greatly influence postemergence herbicide performance. Postemergence herbicides are most effective when applied to small weeds that are actively growing. Application to larger weeds, or plants growing under environmental stress, may result in poor weed control and increased risk of crop injury. Some postemergence herbicides require adjuvants while others do not. Use adjuvants according to herbicide label recommendations. Always consult the label and follow directions concerning application rates, timing, spray additives, application technique, personal protective equipment, and any other restrictions when using pesticides.

IRRIGATION

Soybeans are a relatively deep-rooted crop. In deep well-drained soils with no restricting layers, roots can penetrate to 6 feet. However, as with all crops, most of the roots are concentrated in the upper half of the root zone. Managing a root zone of 3 feet is the general irrigation recommendation.

The crop water use requirement, also know as evapotranspiration or ET, for soybeans ranges from 18 to 24 inches total for the growing season, depending on climatic conditions. Water use is generally higher in the west than the east.

Daily water use varies with the stage of growth and weather conditions. The typical peak water use rate is about 0.32 inches per day, which normally occurs near the beginning of the pod fill stage. Water use is slow at the germination and seedling stages, peaks at or near the full bloom stage, and then declines with maturity. The most critical time for adequate soil water availability is during the end of the reproductive period when pod fill begins. Soybeans produce many flowers relative to the final number of pods, so losing a few flowers to light water stress earlier in the reproductive cycle is not as critical to final productivity as the same water stress during pod fill.

Irrigating once during the single late bloom stage shows a yield advantage as compared to a similar amount of irrigation during the early bloom stage and only a slight increase in yield compared to irrigating both during early and late bloom periods. Scheduling by soil water depletion, which does not limit the total season application amount, shows that though irrigating by 60 percent depletion, uses less water, the yields similar to the 30 percent depletion method. The 30 percent depletion also occasionally had increased lodging. In some years, 60 percent depletion had only one irrigation application.

Medium- and Fine-textured Soils

If the potential root zone is near field capacity of available soil water at planting time, irrigation prior to late bloom or early pod set may not prove beneficial. The available soil water in the top 4 feet of soil will be 8 to 12 inches and should be adequate to supply the plant needs until the late bloom stage if rainfall is near normal. The soybean is quite drought-resistant once it becomes established. Frequently, a single 4-inch application with flood irrigation or a series of smaller applications with sprinklers in the late bloom to early bean-filling stage is all that is needed to supplement natural rainfall. The greater likelihood of rainfall and the slow drainage of these soils can be problems. Irrigation scheduling for these conditions should not allow the soil to saturate, but allow for some reserve soil water storage in case of rainfall. This reduces irrigation costs and prevents drainage problems.

Coarse-textured Soils

Coarse-textured soils (sands) have a much lower water-holding capacity than the finer-textured soils. Soybeans on sandy soils require much more frequent irrigation because of the low water-holding capacity. Irrigation may be required at any stage of growth with the amount and timing adjusted to the depth of rooting and the weather. Soybeans may require 1 to 2 inches of water every 3 to 7 days during the peak use periods.

Water Quality

Soybeans are rated as moderately tolerant to irrigation-water salinity. As a comparison, grain sorghum and corn are rated almost the same (grain sorghum slightly more tolerant, corn slightly less). Wheat is moderately tolerant to tolerant. Alfalfa is rated moderately sensitive. These ratings only serve as a guideline to relative tolerance of crops. Absolute tolerance will vary due to a number of factors including climatic conditions, soil conditions, and cultural practices.

The total concentration of soluble salts in waters can be expressed in terms of the electrical conductivity. This can be measured for the soil water on the irrigation water to be applied. For example, a soil water electrical conductivity of saturation extracts (Ece) of 5.5 millimhos per centimeter would cause a yield reduction of 10 percent in soybeans.

The electrical conductivity of the irrigation water (Ecw) that would cause yield reduction is less than Ece. This is because soil water has a salt load due to contact with soil salts. The salt in the irrigation water will be an addition to the amount already in the soil water. Irrigation water with Ecw of greater than 2 millimhos per centimeter generally starts to enter a high salinity hazard range. The sodium content of the water also is a concern, but it is not addressed here. The local county agent or agricultural consultant can provide additional information on irrigation-water quality.

Soybeans irrigated using poor-quality water through sprinkler systems may be subject to additional damage when the foliage is wetted by the saline water. Salt accumulates on leaves due to evaporation of water from leaf surfaces, causing premature senescence or leaf drop. During the rapid-growth stage, replacement leaves are grown; but late in the season early leaf drop can result in severe yield loss.

Irrigation options for poor-quality water are completing all irrigation before all new leaf growth occurs or changing the irrigation system delivery method. Early cutoff of irrigation conflicts with the general irrigation recommendation of late water applications for best yield especially on sandy soils. This option may be possible on soils with high soil water storage capacity. Several irrigators have attempted to modify their sprinkler packages by equipping sprinkler systems with drag lines. These deliver the water directly to the ground and do not wet the soybean canopy. However, this option must be accompanied by use of appropriate cultural practices to prevent runoff of irrigation water. This may mean use of such activities as maintenance of heavy residue, deep chiseling, furrow diking, or planting in circular rows. These cultural practices increase surface water storage capability and reduce runoff potential. Increasing pivot speed to reduce application depth will also help limit runoff problems.

Crop Rotation

The potential benefit of crop rotation may be a consideration. From an irrigation stand point, irrigators with limited water supplies usually find soybeans compliment corn irrigation since their critical water period is late in the season verses the earlier critical tasseling period for corn. Often both crops benefit from the rotation. Soybeans also can be used as a late season double crop with wheat.

Soybeans are a good irrigated crop option with good yield potential for both full and limited irrigation regimes. Peak daily water use rates for soybeans are about 0.30 inches per day with total seasonal water needs ranging from 18 to 24 inches. Soybeans have an extensive root system and are drought-tolerant. When limited water is available, late season irrigation is generally most important.

SOYBEAN INSECTS

One advantage of growing soybeans is the crop is relatively

free of serious insect problems. A number of different insects do occur in soybean fields, but few are normally of any economic importance, and the damaging species usually are not abundant enough to warrant control measures.

Since the economics of soybean production dictate that every dollar spent for insect control be fully justified, growers should be relatively sure that yield increases are going to more than pay for the application costs. Insect infestation thresholds have been established to indicate when control measures are actually necessary. The cost of control in a locality, the general condition and stage of crop development, and the expected yield and price of the crop should be considered before a final decision on control measures is made. Poorer growing conditions typically mean smaller plants, so fewer insects are needed to cause economic damage. In very severely drought-stressed situations, yield potential (and profits) may already be so low that further expenses cannot be justified.

Foliage-feeding Insects

Insects that feed on the foliage make up the majority of the insects that attack soybeans. Damage by foliage- feeding insects may occur anytime during the growing season, though light feeding will not affect yield since soybean plants apparently have more leaf surface than required for top yields. Soybeans can more effectively compensate for early stand or foliage loss than for late-season stresses.

Pod-feeding Insects

Pod-feeding insects may seriously damage or destroy the beans in the pods. This will lower the yield, especially if infestation occurs late in the blooming period.

Surveying for Insects

It is almost impossible to walk across a field and evaluate the degree of insect infestation. This is particularly true of insects that attack the pods. The plant-shaking method is a useful tool in surveying soybeans especially when damage is being caused by caterpillars. This requires a heavy piece of white or off-white cloth that measures 24 by 42 inches. A l-inch strip of wood 24 inches long may attached to each end of the cloth to make handling easier. plastic-surfaced material (used vinyl tablecloths work well) will help prevent the dislodged insects from escaping before they have been counted.

To begin the survey, pick a random spot in the field, kneel between the two rows, and unroll the cloth from one row over to the opposite row. It is best at this point to slide the cloth forward so it is positioned under undisturbed plants. Place your arms around approximately 18 rowinches of plants on each side of the row and vigorously shake the foliage over the cloth. Most of the insects should fall into the cloth where they can be counted. The number counted in one sampling location represents the number of insects present on 3 feet of row.

A total of 10 random sites should be sampled across the field, and the total numbers of each economic species should be recorded.

Very active insects (grasshoppers) or very tiny insects (thrips) cannot be sampled with a shake-cloth. To estimate grasshopper populations, walk across the field, visualize square yard units ahead, and count the grasshoppers that leave the area as you walk through it. The need for treatment for thrips is based on visible plant injury and relative counts of thrips. Plants heavily attacked by thrips often have lower leaves that turn brown, shrivel, and die. Lower surfaces appear somewhat silvery as a result of the feeding damage. Damage is usually heaviest adjacent leaf veins. Tiny, black spots (waste products) cover the lower sides of the affected leaves.

SOYBEAN DISEASES

Except for charcoal rot and soybean cyst nematode (SCN), soybean diseases normally do not result in major yield losses under growing conditions. From time to time however, weather and growing conditions can combine to produce significant losses from diseases such as Phytophthora root rot, seedling blights, and Phomopsis pod and stem blight.

Disease-free Seed

As with any crop, a healthy soybean field begins with high-quality, disease-free seed. Certified seed, while not guaranteeing freedom from disease, is usually preferable to bin-run seed since, among other things, it is required to meet minimum germination standards. It is particularly important not to save seed from fields infected with pod and stem blight, purple seed stain, bud blight, or soybean mosaic. These diseases can be transmitted either on the surface of or inside the soybean seed.

Crop Rotation

Since many soybean diseases such as pod and stem blight, brown spot, and bacterial blight can overwinter in infested crop debris, rotating soybeans with nonhost crops such as corn, sorghum, or wheat is a sound disease control practice. Other soybean diseases, such as Phytophthora root rot, charcoal rot, and SCN, can persist in the soil for several years making crop rotation less effective. Even in these cases, rotation will help to suppress the disease and limit yield losses.

Seed Treatment

Soybean seed should not be planted when standard tests

show germination is below 85 percent or vigor is low. If such seed must be planted, a seed treatment is usually recommended. For higher quality seed, seed treatmentis not economical unless conditions at planting time are unfavorable for rapid germination and emergence. Soil temperatures lower than 60 degrees Fahrenheit will inhibit soybean growth and promote seed rot and seedling damping-off, particularly if soils remain wet. Other soil factors affecting emergence are soil type, crusting, depth of planting, and herbicide usage. Seed treatment fungicides can be divided into two general categories: contacts and systemics. Contact fungicides will only control those fungi on the outer surface of the seed. Systemic fungicides, on the other hand, are absorbed into the germinating seed and will provide additional protection as the plant emerges. Contact fungicides used as seed treatments include Thiram and Captan. Commonly used systemic fungicides include Vitavax, Agrosol, and Rival. A fungicide specific for the control of Phytophthora and Pythium diseases is Apron. This product is specific to these two diseases and should be applied in combination with other seed treatment products if a broader spectrum of control is needed.

Use higher rates of Rhizobium inoculum if treating seed with fungicide, because prolonged contact with fungicides may damage the inoculum. For best results, treat seed ahead of planting, then inoculate at planting time. Alternatively, both seed treatment and inoculum may be applied at planting.

Foliar Fungicides

Foliar sprays have been shown to improve seed quality in years when warm weather and frequent rains during the reproductive phase of soybean development favors the growth of the pod and stem blight organism. In Iowa, it has been shown that one fungicidal spray at growth stage R6 (pods just visible at one of the four uppermost nodes) was effective in controlling Phomopsis seed decay.

Seedling Diseases

Seedling disease refers to a large group of soil- and seed-borne organisms that attack the soybean seed prior to germination or the plant during the early stages of growth. Symptoms for all of the organisms are similar. They include seed rot, pre- and postemergence damping-off (a constriction of the stem at or below the soil line), and yellowing and stunting of young seedlings. Situations favoring disease development include poor quality seed and adverse growing conditions caused by wet soil, compaction, or cold soil temperatures (less than 60 degrees Fahrenheit). Management practices include using high quality seed, planting when conditions promote rapid emergence, planting at the correct depth, and avoiding herbicide damage. In certain situations, seed treatments may be effective.

Fungal Diseases of Roots and Lower Stem

Charcoal Rot. The fungus causing charcoal rot survives in the soil and directly infects the roots. Severe losses occur frequently on nonirrigated soybeans, especially during hot, droughty periods. Plants turn yellow and die slowly, but leaves remain attached. The symptoms are easily confused with drought stress. Small, black, fruiting structures known as sclerotia form on the lower stem and in the pith. Any management practice that promotes moisture conservation (reduced planting rates, no-till, weed control, irrigation, etc.) will reduce losses. All varieties are susceptible, but late-maturing varieties are more tolerant than early-maturing varieties.

Phytophthora Root Rot. Phytophthora may occur any time during the season if soils remain wet. The fungus survives in soil and produces swimming spores that infect roots. Phytophthora symptoms include pre- and postemergence damping off. Older plants may turn yellow and leaves may wilt. Roots usually appear rotted and a brown discoloration extends from the soil line up into branches and petioles. Conditions encouraging disease include poorly drained soil, flooding from heavy rains, or overirrigation. Resistant or tolerant varieties are the best means of management. There are many races of Phytophthora and varieties should be selected accordingly. Any improvements in drainage also will help. Chemical control is available. Apron seed treatment will protect plants against early infection. Ridomil, applied in the furrow at planting, will provide protection well into the growing season.

Pod and Stem Blight. Yield losses from this disease are infrequent, but seed quality can be greatly reduced. The fungus survives in diseased stubble or is seed-borne. At maturity, symptoms appear as vertical rows of small black spots on the stem or pods. Seeds may be cracked, shriveled, or moldy. The disease is favored by high humidity and rainfall, especially late in the season. The disease can be managed using crop rotation, destruction of infested debris, or chemical treatment. Seed treatments can be used on pod and stem blight infested seed. Foliar sprays may be used on soybeans grown for seed at growth stage R6.

Stem Canker. Stem canker is rarely a problem, but it is easily confused with Phytophthora root rot. The fungus survives in stubble or can be seed-borne. Spores are windblown or water-splashed onto plants. Sunken, brownish cankers with dark brown margins form on lower nodes with green stems above and below the canker. The canker is usually on one side of the stem but can girdle the stem causing the upper portion of the plant to wilt and die.

Fungal Diseases of Foliage, Upper Stems and Pods

Brown Spot. This disease is very common early in the season but seldom persists until maturity, and yield losses are small. It is seed-borne or survives in crop residue. The disease can spread rapidly through a field by the blowing and splashing of spores. Angular brown spots form on leaves. Later, the leaves turn yellow and drop, giving the lower plant a barren appearance. The disease is most common in humid areas where rotation is not practiced. Crop rotation, removal of debris, and planting disease-free seed are recommended management practices.

Downy Mildew. Light infections are frequent. Severe infections are rare but can occasionally result in smaller seed size, low test weight, and poor quality. The fungus can survive on infested leaf debris and on seeds. Spores are spread by wind or rain splashing. Symptoms appear as pale green to light yellow spots. On the lower surface, tufts of grayish mildew develop. Some leaves may yellow and drop. The seed and inner pod surface may be coated with whitish mold without exterior pod symptoms. The disease is favored by high humidity and cool temperatures. Young leaves are most susceptible and late plantings usually have more disease because spores are more abundant. Management is achieved by crop rotation and using resistant varieties. Purple Seed Stain. Seed infection takes place when rain occurs during flowering. There are no foliar symptoms. Seeds are stained with a purplish-brown color, reducing quality. Disease management includes planting disease-free seed, using resistant cultivars, and seed treatment.

Bacterial Diseases

Bacterial Blight. This disease occurs in most fields, but rarely is moisture consistent enough for losses to occur. Bacteria are seed-borne and survive in crop residue. Bacteria are commonly spread during storms containing high winds and heavy rains. Symptoms include small, water-soaked leaf spots that turn brown and have a yellow border. As spots coalesce, the tissue falls out giving the leaves a torn and ragged appearance. Do not save seed from severely infected fields. Crop rotation and avoiding field operations when the foliage is wet are important management tools.

Bacterial Pustule. This disease is very similar to bacterial blight. The key diagnostic feature is the development of raised pustules in the center of lesions on the under side of the leaf. This disease generally develops later in the season when temperatures are warmer. Some resistant cultivars are available, otherwise management is the same as for bacterial blight.

Viral Diseases

Bud Blight. Caused by tobacco ringspot virus, this disease is occasionally present in low incidence. If abundant, do not use the crop for seed. The disease is easily transmitted mechanically. Immature thrips also may vector the disease. Many weed hosts can serve as reservoirs for the disease. Infected plants are dwarfed. The pith in the upper nodes turns brown. Pods may develop brown blotches and seed development may be inhibited. There is often a distinct curling of the uppermost stem tip. Infected plants often remain green until frost occurs. Management of the disease includes using disease-free seed, controlling weeds in and around the field, and roguing of infected plants.

Soybean Mosaic. This is the most common virus disease. The limited frequency of aphid flights may limit the disease. Several aphids (but not the greenbug) transmit the virus. Most infection comes from planting infected seed. Leaves are crinkled, yellowish, and strapped in appearance. The disease is often confused with 2,4-D injury. Plants remain green until frost. The seed is often discolored, especially around the hilum. Management techniques are the same as for bud blight.

Nematode Diseases

Soybean Cyst Nematode. This disease has rapidly spread in recent years. It can be spread by infested soil on equipment or migrating water fowl. Seed grown in infested areas containing soil peds can also spread the disease. Plants are often stunted and yellow with reduced nodulation. Symptoms are easily confused with drought, flooding, herbicide injury, compaction, or nutrient deficiency. Frequently, the only symptom is a gradual production decline in a field over time. Crop rotation and the use of resistant varieties are the keys to management.

Sting Nematode. The nematode feeds at the growing points of the roots, giving them a pruned appearance. The nematode injects a toxin as it feeds, causing the root tips to become necrotic. Above ground, severe stunting is the most frequent symptom. Rotation with a nonhost crop is the primary management recommendation.

HARVESTING SOYBEANS

Field studies in soybeans have shown that 4-bushelper-acre machine losses are not uncommon. Field experiences and research studies have shown machine losses of soybeans can be reduced to less than 1 bushel per acre. The results of a 1989 study conducted at The Ohio State University found the average machine loss in soybeans was 1.4 bushels per acre, but the highest loss was 4.1 bushels per acre. About 40 percent of the operators studied had losses less than 1 bushel per acre, while almost 20 percent had losses exceeding 2 bushels per acre. To reduce losses, combine operators need to know where harvesting losses occur, how to measure losses, what reasonable levels of loss are, and the equipment, adjustments, and operating practices that will help reduce losses.

Soybean losses can occur before harvest as preharvest loss or during harvest as machine loss. Preharvest loss can be high, and it can only be minimized by harvesting in a timely fashion. Machine loss can occur at several places in the combine, but the most common for soybeans is at the header. Cleaning and separating losses are generally minimal compared to gathering loss. Soybeans are fragile and shatter easily, so close attention should be paid to header adjustments and operation.

It is necessary to have a means to enclose an area of 10 square feet that is representative of the entire swath width. For narrow platforms or heads this is probably most easily done by using a continuous strip across the full cutting width of the machine. For large platforms, it may be easier and more suitable to count only in selected areas, such as every other row, across the width of the machine. In any case, the important part is obtaining a representative sample. Rigid frames can be made from heavy wire or rod. Flexible frames can be made from plastic clothesline and wire stakes in the corners.

Loss Measurement Procedure

This procedure is designed to minimize the time and effort required to check harvesting losses. Ground counts are taken only as required.

1. *Determine total loss:* Operate combine under typical operating and field conditions. Stop the combine and back up about 20 feet. In the harvested area behind the combine, count the beans in a 10 square foot area across the entire swath width and enter this count in the loss data table. Divide this number by 40 and enter the loss in bushels per acre. If the total loss is less than 3 percent of yield, keep harvesting. If loss is greater than 3 percent, continue this procedure to pinpoint the source of loss.
2. *Determine preharvest loss:* In the unharvested area in front of the combine, count loose beans on the ground and beans in pods lying loose on the ground in a 10-

square-foot area across the entire width of cut. Enter this count and then divide by 40 to get loss in bushels per acre.

3. *Determine machine loss*: Machine loss is calculated by subtracting the preharvest loss from the total crop loss. If machine loss is less than 3 percent of yield, keep harvesting. If machine loss is more than 3 percent, proceed to check gathering unit losses. 4. Determine gathering unit loss: In a 10-square-foot area across the entire width of cut in front of the combine, count and categorize beans according to the type of loss below. Gathering unit loss is the sum of these four losses.
   - *Shatter loss*: Loose beans on the ground and beans in loose pods on the ground minus the preharvest loss count.
   - *Loose stalk loss*: All beans in pods attached to stalks that were cut but not gathered into the machine.
   - *Lodged stalk loss:* All beans in pods attached to soybean stalks that were lodged and are still attached to the ground.
   - *Stubble loss*: All beans in pods still attached to stubble.
5. Determine cylinder and separation loss: Subtract the gathering unit loss from the machine loss.

Compare harvest loss levels to goal loss levels. Concentrate on machine adjustments and operating practices that will give the least total loss.

Gathering Equipment

Several machinery developments have improved gathering efficiency over that of a conventional cutter-bar platform in soybeans. Some of these are the add-on flexiblefloating cutter bar, the integral flexible-floating cutter bar, the row crop head, and the narrow pitch cutter bar. Tests in Illinois in 30-inch rows showed header losses averaged 8.7 percent of the total yield with an add-on flexible-floating cutter bar, 3.8 percent with an integral flexible-floating cutter bar, and 1.4 percent with a rowcrop head. These tests indicate an advantage of the integral flexible-floating cutter bar over the add-on. The row crop head seems to have a clear advantage over both cutter bar arrangements in these tests.

Tests in Louisiana compared an integral flexiblefloating cutter bar to a row-crop head in 40-inch rows. The average harvest losses were 3.4 percent with the integral flexible-floating cutter bar and 3.0 percent with the row crop head. Additionally, tests were run with the integral flexible-floating cutter bar in 7-inch rows and 30-inch rows to determine the effect of row width on harvest losses. The average losses were 1.4 percent in 7-inch rows, 2.2 percent in 30-inch rows, and 3.4 percent in the 40-inch rows as previously stated. These tests did not show the clear advantage of the row-crop head over the integral flexiblefloating cutter bar that the Illinois tests did in a direct comparison. If the narrow rows are considered, it appears harvest losses would be lower with the integral flexiblefloating cutter bar.

Other tests have shown narrow-pitch (1.5-inch) cutter bars typically have about two-thirds of the gathering losses that occur with the standard 3-inch-pitch cutter bars. Narrow pitch sections and guards can be ordered as optional equipment on some models or obtained from aftermarket companies.

The data given here should indicate representative losses with various gathering units in standing crop conditions. An important factor not addressed here is harvesting under adverse conditions, mainly lodging. Tests have not been done in soybeans to determine losses with different gathering units under various degrees of lodging. Although specialized gathering equipment for soybeans can significantly reduce harvesting losses, there is the cost factor to consider to make sure the equipment is economically justifiable. Some key factors in this decision are acreage farmed, present machinery, and other crops. The flexible-floating cutter bar will work in wheat and grain sorghum. The row-crop head works well in grain sorghum in rows. A comparison can be made to see if the improved gathering efficiency would justify an additional investment in some of the specialized gathering equipment for soybean harvesting.

Combine Adjustment and Operation

Reel speed and position are extremely important in harvesting soybeans. The reel should be positioned for minimum disturbance of standing plants. Reels with pickup fingers will cause the least disturbance to the standing plants. When the reel is positioned properly, the soybean plants will be conveyed smoothly across the cutter bar, along the auger, and into the combine feeder. Gathering losses are minimized when the reel speed is set about 25 to 50 percent faster than ground speed. A 42-inch reel should rotate at about 12 rpm for each 1 mph of ground speed. The reel will shatter beans excessively if it turns too fast, and too many stalks may be dropped or recut if it turns too slowly. Position the reel 8 to 12 inches ahead of the sickle. A bat reel should be operated just low enough to tip the stalks onto the platform.

One of the more important items to check is the separator speed. Each combine has one particular shaft as the starting point for checking operating speed. If the separator is not running at the proper speed, it will reduce the efficiency of the cleaning and separating units. It will be difficult to clean soybeans and minimize losses.

Before the combine goes to the field, there are a number of other adjustments that should be made using the operator's manual as a guide. These include cylinder speed, cylinder-concave clearance, sieve settings, and fan adjustment. If the operator's manual is followed closely, the operator usually needs to make only minor adjustments in the field.

Seed damage is affected more by cylinder speed than by cylinder-concave clearance. It is important to reset the cylinder speed as conditions change throughout the day and the harvest period. Tests have been conducted to compare seed quality with rotary and conventional threshing combines. Percentage of splits with the rotary machines was typically half or less of the splits with the conventional machines.

If weedy conditions are encountered during harvest, especially when weeds are green, reduce travel speed to maintain low threshing and separating losses. Do not increase cylinder speed because this will only cause excessive damage to the beans and put more trash in the grain tank.

DRYING AND STORAGE

Drying Considerations

Soybeans can be harvested at 16 to 18 percent moisture content without damage if the plants have frosted or died. The beans must be dried to less than 14-percent moisture content before storing. Germination is reduced when seed beans dry and rewet several times before harvesting. Drying systems used for corn or grain sorghum are adaptable to soybeans. Soybeans are fragile and can be damaged by air that is too hot or too dry and by rough handling. Three common methods for drying soybeans are natural air, low temperature, and high temperature. Natural air involves moving unheated outdoor air through a bin of soybeans. Only one to two points of moisture can be removed with a natural air drying system. Little natural air drying occurs unless outdoor air temperature is greater than 60 degrees Fahrenheit and relative humidity below 75 percent. The successful use of unheated air depends on the available airflow rate per bushel of beans and favorable weather conditions. If the soybeans can be dried, their quality is excellent unless there is handling damage. Bins should be equipped with a full perforated floor and fan unit. Fans capable of delivering 1 to 2 cubic feet of air per minute per bushel of soybeans (cfm/bu) are required.

Low-temperature drying is similar to natural air. Lowtemperature drying can be used when drying soybeans for seeds and is best suited when only three to four points of moisture are removed. The outdoor air is heated 5 to 20 degrees Fahrenheit with low-temperature drying. This reduces the humidity of the drying about 10 to 20 percent. Bin depth is limited to 16 feet and air flow rates are 1 to 2 cubic feet per minute per bushel. The humidity of the drying air should range from 40 to 70 percent. High-temperature dryers, such as batch dryers, continuous-flow dryers, or in-bin dryers, can be used to dry commercial, those used for oil or meal, but not seed soybeans. It is recommended the maximum drying air temperature be limited to 140 degrees Fahrenheit. If quality (few splits or no change in the oil's free fatty acid content) is important, the drying process should be controlled by maintaining the relative humidity greater than 40 percent rather than controlling temperature. Drying temperatures greater than 110 degrees Fahrenheit and 140 degrees Fahrenheit will affect germination and oil content, respectively. Excessive heat decreases the drying air relative humidity, which causes damage to the seed coat or pericap cracking and results in splits. Seed coat cracking and splits may occur if the relative humidity of the drying air is less than 40 percent. Split soybeans are more susceptible to mold and fungi invasion than whole beans.

Storage Considerations

Beans must be uniformly dried at less than 14 percent if stored less than 6 months and 11 percent for longer storage. Good-quality, clean soybeans can be stored at higher moisture content than damaged beans. The storage moisture content is dependent on the temperature, moisture content, length of storage, and soybean quality.

Storage problems often can be traced to areas where pockets of foreign material, fines, or weed seeds have accumulated. Such pockets provide a place for insects and molds to live, and they inhibit effective aeration and fumigation. Excessive foreign material and fines restrict the airflow through the soybeans. Spoutlines can contain up to 80 percent weed seeds. Germination losses of seed soybeans is more likely due to pockets of excessive splits and foreign material rather than unsafe average storage temperatures and moisture. These pockets of material tend to be at higher temperatures or moisture contents. Molds and fungi are able to grow and invade viable beans in these pockets and kill their germination. These dead seeds are then mixed with viable soybeans, so the average germination is reduced.

Good storage management can greatly influence the storability of soybeans. Information on grain protectants and fumigants and bin wall sprays for use with soybeans can be obtained from the area or state Extension entomologist or county Extension agents. Successful storage of soybeans involves sanitation, aeration, and monitoring (SAM).

Sanitation ensures soybeans are harvested, transported, and handled with clean equipment and stored in a clean structure. Combines, transportation equipment, and conveying equipment should be cleaned of infested soybeans or other grains before harvest. Sanitation in storage structures involves sweeping, cleaning, and removal of old soybeans or grains both inside and outside the storage structure. To avoid contaminating newly harvested beans, remove and destroy leftover debris from bins and sweep down the walls, ceilings, sills, ledges, and floors.

Trash, old machinery, and litter around the bin areas should be removed. No vegetation should be within 2 feet of the storage structure. Another source of infestation is the spilled grain around the loading and unloading equipment. A weatherproof seal or other bin repairs, particularly where side walls join the floor and roof, are needed to prevent moisture or rodents from entering the storage structure.

Soybeans, at 14 percent or less, stored for more than 3 months should be placed in a bin equipped with aeration system. Aeration is reducing the temperatures inside a bin to 40 degrees Fahrenheit using outside air. Aeration controls grain temperature to prevent spoilage, mold and insect activities, and moisture migration. This is accomplished by an aeration system that provides a reasonably uniform airflow of about 0.1 to 0.5 cubic feet of air per minute per bushel or a drying system. Aeration is not a drying process although small moisture changes do occur with a change in temperature.

During aeration (cooling or warming) a temperature zone moves through the beans much like a drying front during drying, only much faster. It is recommended to begin aeration immediately following harvest to remove field heat and equalize temperatures within the storage structure. Aeration fans can be controlled manually or with controllers. The electronic controllers can be used with natural air or low-temperature drying operations and aeration. Simple thermostatic controllers can only be used with aeration. Aeration controllers typically will reduce the average temperatures in a bin 4 to 6 weeks quicker than manual operation. This time difference is dependent upon whether the beans are harvested in early or late fall. Electronic controllers have several management strategies to choose from depending on moisture content, temperature, and storage time.

Soybeans should be maintained at 40 degrees Fahrenheit during the winter and 50 to 60 degrees Fahrenheit during the summer. To cool the beans in the fall, the average outside temperatures (average of high and low for a day) should be 10 to 15 degrees Fahrenheit less than soybean temperatures. The fan should run continuously until the beans are completely cooled unless a controller is used. At 0.1 cubic feet of air per bushel, the fan will have to run about 120 hours to move a cooling cycle through the bin. However, to be sure the beans are cooled, the temperature of the beans should be measured at several points within the bin.

Frequent monitoring or observation is the best way to detect unfavorable storage conditions. The soybeans should be checked weekly from April to November and biweekly from December to March. Operating a fan for about 20 minutes during monitoring helps determine the conditions. Be observant for odors, steam, water vapor, heat, and moisture condensation while the fan is on. A faint musty odor is often the first indication of spoilage. This is particularly true during the spring when heating may be occurring due to warm weather or molds and insects. If steam, water vapor, or heat is observed, the soybean quality has deteriorated.

The aeration fans should be turned on immediately to recool the soybeans and then market them immediately. Fans should be operated during the spring to equalize the temperatures within the bin. The spring target temperature is 50 degrees Fahrenheit. Proper monitoring requires a grain probe, a section of eaves trough or strip of canvas for handling the grain from the probe, screening pans for sifting insects from the grain samples, and a means of measuring temperatures in the grain. A record book of grain temperatures can help in detecting gradual increases in grain temperature. Slight increases are an early sign that heating and potential spoilage may be occurring. Thermometers or temperature monitoring systems can be used to measure the grain temperature.

Storage Facilities

Soybeans can be stored on the farm in round bins or regular concrete-floor buildings. Drying is risky in flat storage because of possible poor air distribution. Beans held in storage, even though dry, need periodic aeration to minimize moisture migration and break up any hot spots that might develop. However, flat storage does have more versatility for other uses.

Beans must be moved very carefully as they are more susceptible to cracking and breakage than most grains. Augers should be operated at low speeds and full capacity to minimize damage. Belt conveyors are better than augers for moving beans. Grain augers are manufactured with bristles on the flighting to reduce the damage during handling. The bristle augers do not work in seed conditioning operations where avoidance of cross contamination of seeds is critical. Handling should be minimized and controlled by limiting drop heights and conveying. Grain spreaders should only be used with commercial beans and avoided when handling seed beans. Bin ladders can be used to reduce the damage due to drop distance during filling.

PROFIT PROSPECTS

Each producer must answer two questions when selecting crops and the acreage of each crop to produce: (1) Will this choice be profitable? (2) Will this add more to the total net income of my farm operation than other choices? That is, is this the most profitable choice? The fixed or overhead costs of land and machinery ownership for soybeans, corn, grain sorghum, and wheat will be basically equal for the production period under consideration. Therefore, the variable costs associated with each crop are the costs that need to be considered when selecting a given crop. Variable costs include labor, seed, herbicide, insecticide, fertilizer, fuel, oil, repairs, crop insurance, drying, custom work, crop consulting, and miscellaneous. Variable costs will vary depending on the management practices used, tillage operations, labor efficiency, and type and fertility of the land. Each producer should develop the variable costs of production for soybeans and any other crop alternatives. Expected yield and selling price need to be determined for each crop alternative.

The decision to produce soybeans will depend primarily on the costs and expected returns for soybeans in comparison with other crop alternatives. However, the producer should take into account other variables such as previous crop rotation, livestock operation, and the machinery and labor requirements of each crop. The type and amount of equipment, crop rotations, and farm size all affect the cost of producing crops. The tillage practices used and their timing also affect yields and production costs. Each producer should compute the expected returns above variable costs for the farm operation as a means of selecting the crops and acreage of each crop to produce. When computing expected returns above variable costs, consider a number of price alternatives.

SUNFLOWER OIL MARKET

Over the past decade, the total demand for sunflower seed, derived from the total demand of sunflower oil, has increased to over 1 million tonnes, which makes South Africa a net importer. Important to note is that it is not the seed or the cake that is imported, but the sunflower crude oil. Table 3 shows that the total consumption of sunflower oil in South Africa is estimated at 451 781 tonnes, of which 321 530 tonnes were produced locally and 130 251 tonnes were imported.

Table 3. Consumption of sunflower oil in South Africa (by segment)

Total sunflower oil market volume by segment - 2000

	Tons
Retail	83 781
General Trade	53 000
Wholesale	110 000
Industrial	52 000
Bulk	40 000
Exports of refined oil (mainly SADC)	113 000
Total consumption	451 781

Although no formal data are available on the consumption of sunflower oil for the period 2001/02, it is estimated that the total consumption decreased because imports of sunflower crude oil1 decreased by 97% as a result of the drastic increase in sunflower crude oil prices and the depreciation of the exchange rate.

Table 4. Usage spread of sunflower oil amoung South African Population

	% Population	*% Expenditure*
African	76.79%	14.08%
White	10.86%	24.88%
Coloured	8.89%	16.67%
Indian	2.47%	33.10%
Other	0.99%	11.27%

It is important to take the usage spread of sunflower oil into consideration. It is interesting to note that, despite the fact that the Indian population makes up only 2.5% of the total SA population, its expenditure on sunflower oil is estimated at a massive 33% of the total SA expenditure on sunflower oil. It is estimated that the white population is responsible for 25% of all expenditure on sunflower oil. It is well known that South Africa is a net importer of oilcake products. The import of sunflower cake shows a decreasing trend compared to the total available oilcakes, whereas the usage of soybean cake increased substantially over the same period.

Over the past five years the total production of sunflower cake has exceeded the total consumption. This has created an increased downward pressure on the price of sunflower cake. The price of sunflower cake is derived from its relative nutritional value compared to other oil cakes. The South African feed industry is capable of utilizing more than 300 000 tonnes of sunflower seed meal per annum. According to the members, usage is restricted by the following criteria:

(i) the ratio of soy-bean meal to sunflower seed meal prices,

(ii) high transport costs to coastal regions,

(iii) inconsistent quality of sunflower seed meal from crushers (high fibre, low protein) and

(iv) the low availability of sunflower seed meal as a result of low oil prices and thus reduced crushing.

Market Structure

Information on the number of sunflower seed producers is not available, but industry sources believe the number of producers is just slightly lower than the number of maize producers. This is due to the fact that farmers plant both crops simultaneously as part of their diversification strategy. The regional distribution remained approximately the same over the two seasons. The Free State is the leading producer with a share of 49% of the total area planted and 52% of the total production, followed by the North-West Province with 36% and 35% respectively. The top eight seed-crushing plants yield more than 300,000 tonnes of sunflower meal annually, with a total

Table 5: The Sunflower seed –to- sunflower oil supply chain

	Units	Jul-03
A) FARMERS		
FARMGATE PRICE	R/ton seed	1576.00
Transport cost: Farm gate to silo	R/ton seed	106.00
Handling & Storage cost: Costs of farmer	R/ton seed	35.00
SAFEX (nearby contract)	R/ton seed	1717.00
B) SUPPLY OF CRUDE OIL		
1) Imports of Crude Oil		
Sunflower Crude Oil, Argentina FOB	US$/ton oil	539.00
Freight	US$/ton oil	40.00
Insurance	US$/ton oil	6.47
Duty US$/ton oil	53.90	
Discharge and clearing transport	US$/ton oil	27.00
Exchange Rate	R/US$	7.53
Sunflower Crude Oil, factory @ Durban	R/ton oil	5017.75
Transport: Durban to Reef	R/ton oil	120.00
SUNFLOWER CRUDE OIL PRICE, FACTORY @ REEF R/ton oil	5137.75	
2) Local supply of Crude Oil: Crushing activity		
Transport cost: Silo to crushing plant	R/ton seed	86.00
Handling cost: Costs of crusher	R/ton seed	35.00
Storage costs: Costs of crusher	R/ton seed	100.00
Interests paid on investment	R/ton seed	140.00
PRICE CRUSHERS PAY FOR SEED	R/ton seed	2078.00
Fixed costs R/ton seed	65.00	
Variable costs	R/ton seed	120.00
Total costs of crushing	R/ton seed	185.00

CAKE PRICE	R/ton cake	900.00	
	Crude oil contribution (39% extraction from 1ton of seed*Oil price)	R/ton oil	2003.72
	Cake contribution (42% cake from 1ton of seed*Seed price)	R/ton cake	378.00
	Costs of seed	R/ton seed	2078.00
	CRUSHING MARGIN	R/ton seed	303.72
	TOTAL CRUSHING COSTS	R/ton seed	185.00
	MANUFACTURERS REALISATION (CRUSHING)	R/ton seed	118.72
C) REFINEMENT	**ACTIVITY**		
	SUNFLOWER CRUDE OIL PRICE, FACTORY @ REEF	R/ton oil	5137.75
	Interests paid on investment	R/ton oil	58.00
	Fixed costs	R/ton oil	188.00
	Variable costs	R/ton oil	180.00
	Total costs of refinement	R/ton oil	426.00
	Total costs before refinement loss	5563.75	
	TOTAL COSTS OF REFINED SUNFLOWER OIL(exl. packaging)	R/ton	5918.88
D) PACKAGING	**ACTIVITY**		
	Interests paid on investment	R/ton oil	65.00
	Fixed costs	R/ton oil	120.00
	Variable costs	R/ton oil	270.00
	Total costs of packaging	R/ton oil	455.00
	Distribution costs	R/ton oil	250.00
	TOTAL COSTS OF SUNFLOWER OIL (incl. packaging)	6623.88	
	MANUFACTURER-TO-RETAIL MARGIN (Pure sunflower oil)	3372.37	
	SUNFLOWER OIL RETAIL PRICE	R/ton oil	9996.25

Table 5. Largest crushing plants in South Africa

Location	*Processor*	*Crushing Capacity (tons)*
Boksburg & Randburg	Nola Industries	400 000
Southdale	Epic	200 000
Lichtenburg	Epko	170 000
Isando Willowton	Oil Mills	100 000
Viljoenskroon	Senwesko	100 000
Pietermaritzburg	Capital Oil Mills	50 000
Isithebe	Elangeni Oil & Cake Mills	30 000
Pietermaritzburg	Sealake Industries	25 000
Total		1 075 000

This implies that, given the current total production of sunflower seed of 706,760 tonnes, only 65% of the total local crushing capacity are utilised. This opens a window for what is in the industry referred to as "toll crushing". Due to the surplus crushing capacity, there is an opportunity for any role player in the industry to crush seed, sell the crude oil at a lower price than the import parity price and still manage to realise some profit. This phenomenon makes the crushing industry highly competitive since the utilisation of crushing capacity is readily available to anyone in the business. The fact that there exists excess capacity in South Africa with respect to the processing of oilseeds is exerting more and more pressure on the ability of large and small processors' to reach and maintain an optimum level of economies of scale.

The South African Oil Processors Association (SAOPA) represents the oil processing industry's interests. Table 6 presents a list of oil refineries, which are currently members of the Association. It is interesting to note that not all crushers have a refining capacity and not all refiners have crushing capacity. Eight of the thirteen refineries are relatively close to Durban harbour from where sunflower crude oil is imported.

Table 6. South African Oil Refineries

Location	*Refiners*
Pietermaritzburg	Capital Oil Mills
Randfontein	Continental Oil Mills
Isithebe	Elangeni Oil & Cake Mills
Southdale	Epic Foods
Lichtenburg	Epko Oil Seed Crushing
Cumberwood	Hentiq 1320
Rivonia	Nedan Oil Mills
Randfontein	Nola Industries
PietermaritzburgSealake Industries	
Isipingo Beach	Sun Oil Refineries
Port Shepstone	Sunola Oil Mills
Durban	UBR
Isando	Willowton Oil Mills

Oilseed processing is highly capital-intensive, and requires specialised knowledge and state-of-the-art technology. The fact that large quantities of crude oil are entering South African harbours for refining makes it very difficult for large and small processors to survive. Oil mills near the respective harbours have a big advantage in this regard, as their transport costs are very low (e.g. only from Durban harbour to Pietermaritzburg). In other words, location of processing facilities and transport cost can be regarded as very important.

Processors, especially large ones, must keep themselves up-to-date with technological innovations in the oilseed processing industry in order to be compete with large overseas processors. This, in fact, hampers the ability of the industry to maintain, or reach, economies of scale. Most of the local crushing plants were established in the mid-eighties and have not been revamped since. The distance from crushing plant to the market also influences role players' ability to achieve economies of scale.

For example, a number of small processors have the disadvantage of being far from the market, in contrast to the larger ones who built their factories closer to the markets of Gauteng and KwaZulu-Natal, or those that are situated near harbours. The issue here is the relative difference in transport costs of transporting the primary commodity to the processing plant and transporting the processed product to the market, which affects, particularly, the smaller processors.

A concern shared by the entire industry is the utilisation of used oils. Used oils comprise between 50,000 and 100,000 tonnes annually that are being recycled back into the economy. Used oils have alarming health risks, a topic that is currently under researched by Professor Kock at the University of the Free State. Used oils are substantially cheaper than virgin oils, and they offer unscrupulous operators the opportunity to exploit consumers. As with the other agricultural sub-sectors, also the oilseeds industry must adhere to several regulations pertaining to food safety. These regulations include, amongst others, the following:

— Foodstuffs, Cosmetic and Disinfectants Act of 1972 (Act 54 of 1972)
— Health Act of 1977 (Act 63 of 1977)
— Fertilisers, Farm Feeds, Agricultural Remedies Act of 1947 (Act 31 of 1947)
— Agricultural Products Standards Act of 1990 (Act 119 of 1990)

Price Formation

As with all other field crops, prices of oilseeds vary substantially from one season to the next. Although sunflower seed is also traded on SAFEX like maize and wheat, the local price of sunflower seed is not only influenced by the supply and demand factors of sunflower seed but also by the supply and demand factors of the local and international sunflower oil market.

The international oil prices act as a guideline for domestic seed and oil prices. In particular, the situation of the Argentinean oil market has a significant impact on the local market since the Argentinean oil market has the same marketing period of sunflower seed as SA producers. Hence, the sunflower crude oil price at the Reef is derived from the Argentinean free on board (fob) price for sunflower crude oil. The crude oil price can be regarded as an import parity price of crude oil at the Reef seen that the costs of transport, discharge and insurance are taken into account.

South Africa is a net importer of oil, which implies that the local price is normally traded close to import parity levels. If a refinery can import crude oil from Argentina at a lower price than that of the locally produced oil supplied by the crushers, the refineries will simply decide to import the oil. This is the reason why more than 60% of all the sunflower oil refineries are close to Durban harbour. South Africa is not a significant role player when it comes to international oilseed production and trade, and, therefore, South Africa is regarded as a price taker.

The sharp increase in the price of sunflower crude oil during November and December 2001 not only was caused by the depreciation in the exchange rate, but it also coincided with a drastic increase in the fob price of Argentinean sunflower oil, which resulted in the opening of a gap between soybean oil and sunflower oil prices. Historically, sunflower oil world prices have traded at an approximate premium of US$20/ton over the world prices of soybean oil. Hence, the fact that the gap between sunflower oil and soybean oil was as high as US$222/ton, during the period December 2001 to January 2002, influenced the behaviour of crushers and refineries significantly. When considering the international market, it becomes clear that not sunflower seed is imported but sunflower crude oil. The price that an oilseed processor is willing to pay for sunflower seed should therefore be derived from the import parity price of sunflower crude oil, as well as from the prices of other oils that can serve as substitutes for sunflower oil, such as soybean oil.

Sunflower Oil Supply Chain

Similar to the maize-to-maize meal value chain analysis, various assumptions had to be made to construct the framework for the unpacking of the sunflower seed–to–sunflower oil supply chain as presented in Table 7. The basic assumptions used in the calculation of this value chain are based on the same principles as the assumptions used in the calculation of the maize-to-maize meal value chain. For this value chain the following five main nodes or levels were identified: the sunflower seed producers (farm level), the crushers of seed, the refineries of crude oil, the wholesalers and retailers and, finally, the consumers.

The farm gate price is derived from the SAFEX average nearby contract price, which is lagged by three months. Statistical tests proved that the level of correlation between the SAFEX price of sunflower seed and the consumer price of sunflower oil is the highest when the SAFEX price is lagged by three months. This implies that it takes three months from when the crushers buy the sunflower seed until the oil appears on the shelf. A number of role players also argued in favour of the inclusion of a 3-month lag period between the SAFEX price and the consumer price simply because of the period of time required for processing the oil as well as the need for basic hedging strategies.

It is important to keep in mind that the value of the crude oil is derived from the import parity price of crude oil. Industry specialists feel that at specific periods in the year the local price of crude oil trades below the import parity price levels. However, it was decided that for the purposes of these calculations a higher price of crude oil will reflect the "worse case" scenario because crude oil is regarded as the raw material and the higher the price of the raw material that enters the value chain, the slighter are the possibilities of making profit if the retail prices are kept at a constant level.

During November and December 2001 the price gap between the world price of sunflower and soybean oil increased significantly as a result of the sharp increase in the price of sunflower crude oil. The manufacturer-to-retail margin became negative in this period. With the soybean crude oil price more than US$ 200/ton cheaper than the crude oil price of sunflower, refineries started substituting sunflower oil with soybean oil and imports of soybean oil increased sharply. The fact that the retail price, which was used for the purpose of these calculations, reflects the price of "cheapest cooking oil" on the shelf, and not pure sunflower oil, implies that it reflects the price of blended oil. Pure sunflower oil is always sold at a premium. This premium can in some stores be as high as 30%.

From 2001 to 2002 sunflower crude oil imports decreased by 97.2%, while the imports of soybean oil increased by 54.9%. Apart from sunflower oil, the imports of all the crude oils have increased. The results suggest that the refineries had to blend different oils into the final product to maintain a positive manufacturer-to-retail margin.

REFERENCES

Pope, Kevin, et.al., "Origin and Environmental Setting of Ancient Agriculture in the Lowlands of Mesoamerica," *Science*, 18 May 2001:Vol. 292. no. 5520, pp. 1370 - 1373.

Shosteck, Robt. 1974. *Flowers and Plants. An International Lexicon with Biographical Notes.* Quadrangle/The New York Times Book Co. 329 pp.

Wood, Marcia. June 2002. "Sunflower Rubber?" *Agricultural Research.* USDA

4

Production of Coconut

The coconut industry provides employment for some 35,000 people in Jamaica. At the end of 1994 there were approximately 8500 registered coconut farmers. The total area under production in December 1993 was 15,957 ha., about 13,202 ha. with trees bearing nuts. These figures represent a great recovery from the 1988 hurricane which caused great destruction on 23,807 ha. of coconut trees.

The Coconut Industry Board (CIB) has been promoting a rapid replanting program to allow the local industry to regain its original economic position and the production of coconut is increasing on the island. Ninety-seven percent of the total coconut produced in Jamaica goes toward household use. Records show that over the past five decades the coconut industry in Jamaica has experienced recurrent hurricanes and many pest problems, including those associated with the coconut mite, *Eriophyes guerreronis*. These problems have resulted in an ongoing effort to keep the industry alive.

The coconut mite is now the major pest of coconut in Jamaica and is spreading rapidly throughout the island. Its feed injury retards fruit development by causing the development of broken striations of the exocarp which arrest subsequent growth at the point of attack. The coconut mite has proven to be difficult to control. A wide range of chemicals has been employed to control it over the past two decades but the results have not been satisfactory. Good plant husbandry has been recommended to alleviate the economic impact of the mite on coconut production. In the mean time, research has been directed toward the identification of resistant coconut varieties and biological control agents.

Although the coconut mite is an international pest, its biology has not been researched to any great extent. Keifer *et al.* amented the "surprising" neglect of research in eriophyid systematics "considering the group's economic importance". They also noted that literature on eriophyids is "generally scattered and fragmented". There is urgent need for comprehensive research in all areas of biology, economic importance and management of eriophyids, particularly, *E. guerreronis*.

The objectives of this thesis were

1) to review the available literature regarding the coconut mite, its host, and their interaction;
2) to survey the distribution and extent of damage caused by the coconut mite in two of the three main coconut growing areas in Jamaica;
3) to describe the seasonal variations in coconut mite damage on Maypan and Red Malayan Dwarf coconut varieties;
4) to compare the population fluctuations of the coconut mite on Maypan and Red Malayan Dwarf coconut varieties; and
5) to evaluate the yield loss due to coconut mite damage on selected farms in eastern Jamaica.

There is a great paucity of information on the biology and management of the coconut mite, and the information that exists is incomplete. The present review was written to organise the available literature on the coconut mite and the coconut industry.

COCONUT TREE

Cocos nucifera Beccari or coconut belongs to the palm family, Arecaceae which consists of 200 genera and over 2,000 described species. According to Woodroof the term coconut is derived from the Spanish and Portuguese word, "coco", which means "monkey/grotesque face", but the plant is known in many countries by local names. For example, it has been known as "naryal" in India for millennia and as "nut of India" by Cosmos, the Egyptian traveler, in AD 545.

The tree itself has been described as, "man's most useful tree", "king of the tropical forest", "tree of life", "tree of heaven" and lazyman's crop, *inter alia*. Coconut varieties fall under two broad groups, Tall or *typica* and Dwarf or *nana*. Tall and Dwarf coconut types may hybridise to produce intermediate forms. The Tall variety has greater genetic variability as it is usually cross pollinated.

The coconut plant is monoecious, producing both male and female flowers. The male flowers are located distally while the female flowers are found proximally on each inflorescence. The type of pollination is determined by the relative maturation times of the male and female flowers.

In the Tall varieties the male flowers open before the female flowers, hindering self pollination while, an overlap of the opening phases of male and female flowers in Dwarf plants allows for self pollination and greater tendency toward homosygosity. Coconuts are also named after areas where they are grown long enough to have developed distinctive characteristics, e.g., Panama Tall, West African Tall and Malayan Dwarf. Maypan variety is a hybrid of the Malayan Dwarf and Panama Tall varieties.

The traditional commercial coconuts were the Tall varieties which were preferred above the Dwarf varieties because of the quality and quantity of copra they produce. They normally live for over 60 years, are adaptable to a wide range of soil conditions, fairly resistant to diseases and water stress, and start to bear within six to ten years. The Dwarf varieties come into within three to four years, attain full production by the ninth year and have a life span of about 30 to 40 years.

While they show greater susceptibility to some diseases, the Dwarf varieties exhibit greater resistance than the Talls to some viral diseases, including lethal yellowing. Each coconut inflorescence emerges from the base of a leaf and is approximately 120° around from the previous one. After fertilization of the female flowers, each inflorescence develops into a cluster of fruits called a bunch.

Occasionally the spikelet of an inflorescence is in direct contact with the spikelet remnants of an older bunch. The native habitat for the coconut palm is unknown because coconut is dispersed by water, although human activity could be credited for much of its dispersal. Of all the cultivated trees in the world, the coconut palm has the widest geographical range. Ninety percent of the world's coconut acreage lies within 20° N and 20° S of the Equator.

The crop is best grown within 600 ft. above sea level, with over 1,250 mm rainfall or a high water table on a rich silty loam. Coconut was introduced into the West Indies at the beginning of the 20th Century. The main variety was the Panama Tall found in Guyana, Jamaica, St. Lucia, Venezuela, Trinidad and Tobago. There was also the dwarf variety, with its Yellow, Red (golden), and Green colour morphs.

Importance of Coconut

The coconut palm and its fruit are regarded as the most important plant to humans around the world. Among its most important uses coconut is a food source, provides supplement for body fluids and minerals, and acts as an antihelminthic. The liquid endosperm is also a media for *in vitro* storage of semen and a growth regulator of plants. Copra, the dehydrated endospern of the nut, is next to soybean as a source of oil for food. Coconut oil is also used in cosmetics and pharmaceuticals.

Table 1. Estimated world production of coconut, 1989-1993 (in metric tons copra equivalent)

Country/Region	*1989*	*1990*	*1991*	*1992*	*1993*
APCC1 Countries	6,938,582	7,972,720	7,250,668	7,800,361	7,934,896
F.S. Micronesia	8,000	7,500	7,000	6,500	6,500
Fiji	24,769	33,505	27,048	32,500	33,000
India	1,275,000	1,397,000	1,448,000	1,504,000	1,698,000
Indonesia	2,221,000	2,332,000	2,337,000	2,511,000	2,515,000
Malaysia	199,000	224,506	182,721	209,389	189,759
Papau New Guinea	186,000	164,000	127,000	163,000	176,000
Philippines	1,876,000	2,629,000	2,060,000	2,238,000	2,197,000
Solomon Islands	42,413	39,209	28,899	36,482	36,637
Sri Lanka	504,000	514,000	443,000	466,000	439,000
Thailand	345,000	342,000	331,000	331,000	338,000
Vanuatu	42,000	62,000	46,000	44,000	44,000
Vietnam	173,400	190,000	180,000	224,490	222,000
Western Samoa	28,000	24,000	19,000	20,000	26,000
Palau	14,000	14,000	14,000	14,000	14,000
Non-APCC Countries	1,370,250	1,329,000	1,334,574	1,363,236	1,340,484
Asia	98,750	97,250	97,584	107,769	105,106
Pacifics	66,750	64,250	63,775	65,075	63,375
Africa	488,500	461,250	441,552	428,534	429,677
C. America/ Caribbean	450,500	421,000	405,898	438,900	407,164
South America	265,750	285,250	325,765	322,958	335,162
Total	8,308,832	9,301,720	8,585,242	9,163,597	9,275,380

The material that remains after the oil is expressed from copra is called oilcake and is used as animal feed. Coconut shell is used directly as fuel, filler, extender in the synthesis of plastic, to make activated charcoal, household articles, and to produce various distillation products, such as tar, woodspirit and pitch. Coir, a course fiber from the husk of the nut, has various domestic and industrial uses. Coconut root is brewed and used in folk medicine, for example, as a cure for dysentery.

Coconut accounts for a large part of the national earnings of the Asian and Pacific Coconut Communities (APCC) countries. In 1993 the APCC account for 86% of the world's coconut production. Indonesia, the Philippines, and India accounted for 81% of the coconut produced by the fourteen APCC countries in 1993 and 69% of the World's total.

In Jamaica, coconut production has fluctuated over the last 50 years, due to several hurricanes and lethal yellowing disease. Maximum copra production was 19,245 metric tons in 1971 but was reduced to 8,540 in 1974 and to insignificant amounts in the late 1970's. Coconut production in Jamaica began to recover in the 1980's. Today over 90% of the coconut produced goes toward domestic use and copra is of secondary importance.

Pests

Besides pathogens, there are several invertebrates and vertebrates which attack various stages of the plant. At least 750 insect pests of coconut have been recorded from around the world. These pests attack the leaves, stems, flowers, nuts and roots of the coconut plant. In the Caribbean over 26 major pests have been recorded. Among them are two species of mites, 15 species of insects and three species of rodents.

COCONUT MITE

Of all the coconut pests, the coconut mite is the most important one in Jamaica and the Caribbean. The very first record of coconut mite injury was in 1928 in New Guinea. The injury was then believed to be due to a disease caused by a complex of fungi and bacteria. In the early 1960's the coconut mite was discovered and the family identified from samples of coconut in Guerrero, Mexico and Santa Marta, Colombia during independent research.

In 1965 the coconut mite was described and identified as *Aceria guerreronis* Keifer and in 1971 reclassified by Newkirk and Keifer as *Eriophyes guerreronis* (Keifer). The coconut mite first achieved pest status in 1960 in Mexico followed by Venezuela, Columbia, Trinidad; Puerto Rico, St. Vincent, Grenada and St. Lucia and Virgin Islands. Damage was also seen in several parts of West Africa including Benin, Cameroon, and Nigeria. Griffith attributed this rapid spread of the coconut mite to its ability to take advantage of long distance dispersal by wind currents.

In Jamaica, the coconut mite was first recorded in 1941, in St. Ann parish, but not was recognised as a pest until 1972. The Jamaica coconut industry suggested that an extended period of drought might have led to an outbreak of the mite. Mariau and Otterbein noted that climate affects the development of coconut mite populations. Coconut plants nearer the Atlantic Ocean showed less coconut mite damage.

In Benin and the Ivory Coast, Julia and Mariau found levels of attack up to five times higher in the wet seasons than in the dry; the reverse was reported from Guerrero, Mexico. Otterbein reported that in Costa Rica the greatest nut damage was associated with frequent heavy rainfall and high humidity. Howard *et al.* found that the coconut mite populations increased immediately after periods of high rainfall in Puerto Rico and Florida but noted that coconut mite populations fluctuations were not associated with dry and wet seasons nor with mean daily temperatures.

Mite Populations

The coconut mite breeds and feeds under the perianth of coconut fruit and is most active outside the perianth during late nights and early mornings. Colonisation of coconuts by coconut mites normally takes place within one to six months after fertilisation. Fertilisation takes place within the second month of flowering. Coconut mite populations peak on 3- to 6-month old nuts, after which, the numbers decline sharply. Thus, nuts over nine months old have relatively low populations.

Mite populations are aggregated so that peak densities may exceed 1500 mites/cm^2 and may reach about 4600 mites per (Malayan dwarf) nut, about 3 to 4 months old. Coconut mites leave nuts two to three months before the nuts are fully developed or when damage to the pericarp exceeds 15% of the total surface area because there is no renewal of meristematic tissues. In addition, damaged nut surfaces tend to secrete resin which traps and kills the mites.

Migration may also be density dependent and dispersal may be short ranged or long ranged. Short range dispersal is aided by water and insects. Rain may wash off mites on to nearby open flowers. Coconut flowers are cross-pollinated by insects, particularly Hymenoptera, which also help to disperse coconut mites. Coconut mites may crawl from infested nuts onto uninfested nuts where these nuts are in contact with each other. Wind currents are the most important means of long range dispersal. This is particularly so in the dry season when populations are high.

Losses in Copra

Losses in copra yields ranged from 10% in Benin, 16% in the Ivory Coast, 20-30% in St. Lucia, 25% in Grenada and 30-80% in different areas of Mexico. Julia and Mariau and

Moore *et al.* found copra yield to decline with increasing severity of damage caused by the mite. Damage to the pericarp was categorised through a visual estimation method by Mariau and Julia and was modified by Moore *et al.*

Coconut Mite Management

Chemical control: Generally, chemical controls are either ineffective or impractical and the current emphasis is on biological and cultural control methods. Mariau carried out the first experiments in the Ivory Coast to test the efficacy of 24 insecticides and acaricides on the coconut mite. The four most effective chemicals were, cyhexatin, chinomethionate (morestan), endosulfan and monocrotophos (nuvacron). Successful control was only achieved after up to six applications of these chemicals per year. Similar results were obtained by Mariau and Tchibozo, Hernandez and Julia and Mariau.

However, Otterbein did not observe any significant control of the mite with these chemicals. Griffith noted that stem injection of systemic vamidithion was effective against the coconut mite in Brazil and in Trinidad. However, this method of control is traumatic for the coconut trees and impractical for pure stands on large plantations. Moore *et al.* also argued that these systemic pesticides posed a threat to human health, especially where jelly or water coconuts are heavily consumed.

Environmental control: Petrolatum coatings or polybutene stickers, with or without acaricides did not impede the migration of the coconut mite onto coconuts. Alam suggested that field sanitation, such as the removal of mite infested nuts and dry leaves, and thinning of nut population in high fruiting trees could reduce coconut mite populations. Proper drainage, weed control, adequate fertilisation and replacement or rehabilitation of old plants for controlling the coconut mite have also been advocated.

Moore *et al.* found increasing levels of potassium and decreasing nitrogen levels to be associated with decreasing mite damage. While some of these cultural practices, including increasing soil water retention, might not directly impact coconut mite populations, they improve the health of the trees and ultimately, their tolerance to damage by the coconut mite. For example, Mariau found copra loss to decline with irrigation. He suggested that during periods of moisture stress, nut growth is slower and so the meristematic tissue is subjected to extensive damage by the coconut mite.

Moore *et al.* also proposed that greater yields could be achieved through improved agronomic practices, combined with resistant varieties. Sarangamath *et al.* concluded that copra yields were dependent on various factors, including, variety, age of palm, soil, climate of the area, maturity of the nuts, seasons of harvest and period of storage.

Varietal resistance: Varietal differences in the susceptibility of coconut plants to infestation by the coconut mite have been reported. The shape and colour of coconuts are the two main characteristics that determine the susceptibility. Dark green nuts are thought to be resistant to the coconut mite. A strong correlation was found between the roundness of coconuts and damage by the coconut mite.

Rounded morphology causes greater adpression of the perianth onto the nut surface, retarding entry of the coconut mite beneath the perianth and the mite's ability to establish a colony on the nut. Moore also found two types of bract arrangement on coconuts that offer different degrees of adpression. The degree of adpression decreases with increasing size of the nut.

Biological control: Various arthropod predators were found on the perianth and surface of the coconuts but are not considered to have a significant regulatory influence on mite populations. Entomopathogenic fungi are likely to play a significant role in the control of coconut mite populations in the future. Tests conducted to determine the efficacy of *Verticillium lecanii* in regulating the coconut mite population in St. Lucia showed promise. The entomopathogenic fungus, *Hirsutella thompsonii* Fisher was recorded on the coconut mite in Mexico, Jamaica and Ivory Coast. *H. nodulosa* Petch was recorded on the coconut mite in Cuba. The role of these fungi as control agents is being explored.

COCONUT RESEARCH AND DEVELOPMENT

The South Pacific Island nations are collectively one of the major producers and exporters of copra. Copra and coconut products continue to be an important for the domestic and export markets. Coconuts are important for household food security and the many uses of coconuts that make them an essential part of life in the Pacific. Mainly small holders produce coconuts and the decline in copra prices is increasing poverty in the Pacific.

The environmental benefits of coconuts are very important in the fragile ecologies of the Pacific Islands. These benefits include prevention of soil and beach erosion, nutrient recycling, windbreaks to prevent wind erosion, provide shade to reduce soil temperatures and conserve soil moisture. However the coconut industry is undergoing rapid change due to competition from other edible oils and more profitable crops. The coconut paradigm must change with the times and that is the challenge that faces us today. Despite the decline in copra prices coconut remains an important component of agriculture in the Pacific.

Cogent in collaboration with international, regional, and national partners have made a significant contribution to the coconut industry in the South Pacific. A number of coconut research and development strategies have emerged in discussion among the Pacific Islands that will build on previous projects and activities and provide the foundation for the increase in income for small scale coconut farmers in the Pacific through immediate development activities and long term research.

The potential components for a regional coconut project consist of the following:

— Value adding
— Coconut-based farming systems
— Germplasm conservation and exchange
— Indigenous coconut germplasm
— Hybrid trials
— Synthetic varieties

PROJECT COMPONENTS

In order to increase the income of coconut farmers, there is a need to have more options on what to do with coconut for food and non- food products. There is widespread interest and support for this component of the project. In discussions thus far the greatest area of interest is for technologies for oil extraction in a variety of sizes and costs for families or groups.

There are coconut products and technologies that are indigenous in the Pacific and some new technologies from Asia that may be appropriate for the Pacific. The primary focus is on coconut value adding for household use and the local market including the tourist market. Aspects of the valueadding component include:

— Pilot projects at the village level in each of the participating countries to document the current value adding that is being done in coconut producing areas, determine appropriate new technologies and products, and introduce new technologies and products along with training.

— A national and region inventory of the various kinds of current and potential coconut value adding that could be done at the village level. Some traditional coconut technologies are disappearing and there need to be documentation of these technologies at the national level. A regional inventory of coconut technologies and products will encourage the sharing among Pacific Island countries.

The suggested steps in the process:

1. Use Participatory methodologies at the village level to determine what coconut products rural families are producing. This would be a pilot effort looking at several villages where coconut production is or was one of the important activities.
2. At the same time determine new products and technologies with market potential that could be promoted in the village and assess their interest for trying new technologies.
3. Provide training and test technologies with the villages.
4. Follow up assessment of the technologies and assist with problems.
5. From these pilot studies it should give us an indication of the direction to pursue for wider national efforts in value adding that is appropriate for Pacific islands.

The project will provide funding assistance for carrying out the village assessment and village training, training of extension staff in coconut products and technologies and costs of some coconut technologies. Information on technologies will be documented at a national level and the inventory of technologies and products will be compiled at the regional level by SPC.

Coconut-based Farming Systems

Coconut-based farming systems are important to the

present and future of the coconut industry. There is a wide range of experiences in the Pacific and from other regions concerning coconut based farming systems. Information needs to be compiled so that the various systems are documented, economic data collected, and labour requirement are described. This component will build on the information that was gathered during the Cogent/IFAD project. As a result a publication will be produced with information on coconut based farming systems that can be used by extension staff with farmers under a variety of conditions.

The project will provide assist the countries to gather the information in a standardised form, compile the results, reviewing the literature and developing a publication.

Germplasm Conservation

The International Coconut Genebank in PNG and laboratory are established and operational. Coconut embryos have been sent to the ICG from the SPC Regional Germplasm Centre that had been collected in Cook Islands, Kiribati, Marshall Islands and Tuvalu by Jean-Pierre Labouisse and Dr. Roland Bourdeix of CIRAD with the assistance of national agriculture staff.

Coconut embryo training was done with Fiji, Samoa, and Tonga earlier this year. Recently in collaboration with Fiji Ministry of Agriculture some shipment of Fiji Tall embryos have been sent to the ICG. Many lessons have been learned about transferring coconut embryos so that it can be done effectively.

The regional project will provide support to transfer of coconuts to the ICG, support to the ICG laboratory and for establishment of the field ICG and movement of coconut embryos from the ICG and other sites to the Cogentmember countries.

Indigenous Coconut Varieties

There is a need for further pre-prospecting and collection of other important farmer varieties. Different coconut cultivars have different uses from farmer's perspective. These populations are usually not collected because previous collection activities concentrated on copra yield. Some of the countries in the region have started this work in the ADB/Cogent phase 1 and 2. The target population should also include those with special adaptation to extreme agro-ecological conditions like atoll, highlands and also swampy lands.

In this work, to determine whether the populations are really unique the following activities need to be carried out: a. Morphological characterisation as stipulated in the Stantech Manual and complimented by DNA finger printing. Cirad and Cogent have standardised the use of such analysis and the capability to do the analysis should be developed in the Pacific. Training of staff of the PIC's would be required also for the collection of the sample and it's safe transfer to the laboratory. Those selected from the project would be conserved *in situ*.

The project will provide training, some collecting costs, technical assistance and assist in preparing a laboratory in the region for DNA analysis capability.

Hybrid Coconuts

A project called "Production and Dissemination of Improved Coconut Cultivars" (PDICC) was started in 1989 as a component of the EU funded Pacific Regional Agricultural Programme (PRAP). The purpose of PDICC Project was the improvement of the potential of coconut production in the region by increasing the choice of hybrid coconut cultivars in the South Pacific Region through the evaluation of diverse hybrids in at the Saraoutou research station now the Vanuatu Agriculture Research and Training Centre (VARTC).

During the Project, 39 new hybrid crossings were created by hand-pollination and 8 trials were successfully established in Saraoutou station between 1992 and 1999. This represents a total surface area of 57 hectares with approximately 9000 palms under individual observation. Each of first seven trials incorporates hybrids created by crossing diverse dwarf cultivars with a tall cultivar native of different countries of the region.

The results of these trials will permit the construction of a database of great value for the research and extension services within the region. To date, a wealth of data has been collected and analysed on the characteristics of these hybrids including: speed of germination, age of flowering, production, fruit component analysis and oil content. As two cyclones struck VARTC in 1999 and 2001, interesting observations were done on the resistance of each hybrid to strong winds.

- 4 years after planting, all the hybrids bear flowers.
- There is significant difference between the hybrids for the resistance to strong winds. Malayan Red Dwarf x Rennell Tall (MRDxRIT) is the most susceptible.
- For the yield, the hybrid MRDxRIT shows good performances in all the trials. The hybrids Malayan Yellow Dwarf x Rennell Tall, Brazilian Green Dwarf x Rennell Tall, Malayan Red Dwarf x Tonga Tall, Malayan Red Dwarf x Rotuman Tall seem also very promising.
- All the hybrids with Madang Brown Dwarf as female parent show the higher copra content in the trials. The hybrids with Brazilian Green Dwarf as female parent and the hybrids with Rotuman Tall as male parent have also good copra content.
- For oil content, there are slight differences between the hybrids. MRDxRIT shows the lower oil content.

Full evaluation of a hybrid cultivar takes 9-10 years after field planting is completed. So the evaluation of the first 7 PRAP trials established in Vanuatu will be achieved by Year 2005-2006. Until then, they must be maintained and the yield and characteristics of the fruits must continue to be recorded and analysed.

The project will provide funding to VARTC, which is now under the management of Vanuatu, to continue the trials and gather data so that is can be used in coconut breeding programmes in the future.

Synthetic Varieties

There is a need to increase the productivity of coconuts. One of the ways of doing this is through hybrid production; which is also commonly practiced. While this is effective (hybrid vigor) it is however expensive to establish and maintain seed gardens and also offspring of the resulting original cross cannot be used as seed material because the genetic vigor is not maintained. Also hybrids do narrow down the diversity.

One way of that has been suggested to overcoming these problems is the use of synthetic varieties, whereby 5 to 7 high yielding and adapted varieties are interplanted in a block of 15 hectares. The resulting offspring could be also used as seeds while having an increased yield. The varieties used would have to have synchronised pollination and vigorous selection in the seed nursery.

There would also needs to be adequate isolation to prevent foreign pollen from entering the block. This method has been successfully used in the Philippines it is proven to yield 2-3 tones per hectare. To assist in this program Cirad and VARTC have carried out a number of evaluations on a number of coconut cultivars both from the region and also from other countries. This would assist in selection of the germplasm that would be used for creating the synthetic varieties.

The project will supply some initial funding for the establishment of pilot sites at the ICG in Papua New Guinea to test this approach. This would be a long-term project that would require substantial commitment by the host country. The CRDSP project will have duration of 3-5 years. SPC will coordinate the project through the SPC Agriculture Adviser and PGR Adviser.

A part-time project leader from the Pacific will focus on the various regional outputs of the project on value adding, coconut based farming systems and indigenous germplasm. Technical advice and short-term training can be provided from member countries, Ciard, or through Cogent member countries.

REFERENCES

Julia, J.F., and D. Mariau. 1979. New research on the coconut mite *E. guerreronis* (K), in the Ivory Coast. Oléagineux (France) 34:181-189.

Keifer, H.H., E.W. Baker, T. Kono, M. Delfinado, and W.E. Styer. 1982. An illustrated guide to plant abnormalities caused by eriophyid mites in North America. U.S.D.A. Agric. Res. Services. Agriculture Handbook No. 573:5-121.

Persley, G.J. 1992. *Replanting the tree of life: Towards an international agenda for coconut palm research.* Redwood Press Ltd. Melksham 156 p.

Medina-Gaud, S., and E. Abreu. 1986. "The coconut palm eriophyid mite", *E. guerreronis* (Keifer) in Puerto Rico (Acarina: Eriophyidae). J. Agric. Univ. P.R. 70:223-224.

Moore, D. 1986. "Bract arrangement in the coconut fruit in relation to attack by the Coconut Mite *E. guerreronis* Keifer". *Trop. Agric.* 63:285-288

Woodroof, J.G. 1970. *Coconuts: Production, processing, products.* The AVI Publishing Co., Inc. 241 p.

5

Jatropha Oilseed Production

Among the species producing oilseeds, Jatropha (*Jatropha curcas*) is the frontrunner as a species for commercial cultivation because of its several advantages over others in this category. In the wake of the interest being generated about Jatropha, a large number of people have entered the scene as stakeholders. Unfortunately, not many of them are well informed about its actual potential as an oilseed bearer for biodiesel production.

The current situation is one where overenthusiastic or unscrupulous promoters are misleading a majority of the prospective Jatropha producers. Tall claims are being made of varieties and vegetative planting material that start producing seed in the first year itself, and have the potential to eventually yield 10 tons of Jatropha seed per ha annually. In reality, however, there does not appear to be a single plantation in India that can substantiate such claims.

BAIF Development Research Foundation in Pune has been conducting research on the cultivation aspects of Jatropha for the past six years on its farm in Tiptur in Karnataka. It also organised a National Workshop on the subject in 2003. Since then, BAIF has been serving as an information network centre for those interested in tree-borne oilseed species.

JATROPHA

Apart from Jatropha, the category of tree-borne oilseed species includes Karanj (*Pongamia pinnata*), Mahua (*Madhuca latifolia and M. indica*), Undi (*Calophyllum inophyllum*), Neem (*Azadirachta indica*) and Simarouba (*Simarouba glauca*). Jatropha has an advantage over others in many ways. It is at a higher stage of domestication than others and hence considerable research has been carried out already to understand the species and its cultivation.

Unlike most other perennial species, Jatropha has a shorter gestation period and regular seed harvests are possible within four years of establishment. Being a small tree with a lax canopy, it is ideally suited for small farm agroforestry systems. Its natural distribution throughout India is indicative of its adaptation to diverse agro-climatic conditions. Moreover, farmers are familiar with Jatropha as a hardy fence plant that can survive with very little inputs or management. These advantages notwithstanding, it should not be assumed that Jatropha can succeed under any condition.

Another advantage of Jatropha is the properties of its oil for biodiesel production. Its very low Free Fatty Acid content, almost similar to edible oils, makes it ideal for transesterification and the oil to biodiesel conversion ratio is higher than in other non-edible oils. A higher ratio means better profit margins and hence Jatropha oil is a more desirable raw material than the oils of other tree seeds. Combustion studies have also shown Jatropha biodiesel to be superior to others as its emissions are less polluting than that of others.

Common Beliefs

Whenever a crop species or agricultural practice is in the limelight, a lot of information is circulated to highlight its merits. Not all of it is necessarily based on hard evidence. Common beliefs, distorted claims and wishful thinking are all part of the information dissemination that Jatropha is going through at present.

Hardy species

The commercial success of any species depends on its potential (as a plant) getting realised. The experiences of many who make unrealistic claims are based on observations of Jatropha as a plant and not in a cropping situation. It is common knowledge that Jatropha is a hardy species that survives under extremely harsh environmental conditions with little or no external inputs and management. But mere survival is not what is expected of a crop when it is cultivated commercially. Moreover, in most farms where it is found, Jatropha is planted as a single row fence plant or on bunds. Performance of a species in terms of growth, susceptibility to pests or diseases and seed yield in block plantations will not be the same as that obtained in linear plantings.

Planting material

The upsurge in interest in Jatropha has resulted in the growing demand for its planting material. A natural extension of this demand is the emergence of planting material suppliers who claim to have elite varieties. It is not known whether these so-called varieties have undergone any performance testing under diverse agro-climatic conditions. Studies conducted by BAIF in Karnataka showed pronounced differences among genotypes and the genotype collected locally outperformed that from Maharashtra and Nicaragua.

However, in the absence of multi-locational evaluation, it cannot be concluded that this genotype is superior to others. Seed oil content is another parameter where inflated claims are made. Generally, oil content of Jatropha kernel is around 25%, but some researchers have reported twice as much in seed collections from natural stands. It is too early to say whether such high oil contents can be achieved in large-scale cultivation. Moreover, if the trait of seed oil content is not combined with seed-yielding ability, it is only of academic interest.

Another trend is to encourage the use of vegetative propagules of Jatropha as planting material. There are nurseries that sell propagules at exorbitant rates with the promise that they will come into seed production early. It is an established fact that propagules from cuttings or grafts start bearing early and show uniform growth. Vegetative propagation has been very successful in plantation crops like tea and rubber, but the recommended clones of those species have undergone testing for long periods under biotic and abiotic stresses. The suitability of vegetative propagation for Jatropha is untested. Until studies are conducted on this aspect, Jatropha seedlings that have a tap root system may be a safer option for wastelands than propagules with adventitious roots.

Yield

Enthusiastic promoters of Jatropha often claim that a seed yield can be had in the first year itself, and the yield can be as high as 8.0-10 tons per ha per year when the trees mature in 3-4 years. These are highly optimistic projections. Although it is a perennial, Jatropha starts flowering in the first year itself and may produce some seeds.

Harvesting these seeds is undesirable because the priority for the first three years of establishment of a commercial plantation is to develop the tree structure. This is achieved by periodic pruning and thinning of branches. During this juvenile phase, flowers that appear should be removed at the bud stage so that vegetative growth continues unaffected.

Another over-estimation is the yield projection based on the yield of a single tree or few trees. For example, a vigorously growing six-year old Jatropha tree growing on a bund in BAIF's Karnataka farm produced more than 3.0 kg of seed. It will be unrealistic, however, if the per ha seed production is estimated based on the unusually high seed yield of this single tree.

Inputs

The belief that Jatropha grows with little or no inputs is also a generalised one that does not hold good for commercial cultivation. Repeated harvest of seed in substantial quantities will drain the soil as well as the plant of nutrients and moisture. Unless these are replenished through fertiliser or manure application and irrigation, seed production cannot continue. If the goal is to produce seed on a commercial scale, input application has to match the crop requirements. Such requirements are usually much higher for trees grown for the purpose of production than those serving the function of live fencing.

Multipurpose species

Jatropha has several other uses besides the seed oil. Almost all its parts have medicinal properties and are used in native medicinal preparations. Extracts of its bark can be made into a wax. Tannin or dye can be extracted from leaf and root of Jatropha. These characteristics are projected as additional benefits from the multipurpose Jatropha tree. In reality, if plant parts are harvested for other purposes, seed yield will be adversely affected. Therefore, commercial cultivation should focus on the seed and the periodically pruned branches and fallen leaf litter should be considered as by-products.

Commercial Cultivation

The manner in which Jatropha is promoted at present with exaggerated claims is fraught with the possibility of growers losing interest in this species if their very high expectations are not met. This has happened in several other crops in the past. It is not that these crops failed altogether, but their potential was blown out of proportion and farmers were unhappy with what they eventually realised. Even Jatropha went through such an experience when farmers discontinued its commercial cultivation in parts of Maharashtra a few years ago. However, the current situation and the purpose for which it is promoted now are totally different, so the chances of its success are greater this time around. The potential seed yield of Jatropha, according to technical literature, is up to 10 tons per ha.

In India, apparently, the recorded yields have been less than 2.0 tons per ha. BAIF investigated seed production of three genotypes of Jatropha under three plant densities in Tiptur district of Karnataka. This aim of this study was to examine the productivity of Jatropha under average resource conditions. Therefore, it was conducted on a marginal land with limited fertilisation and management. The field received life-saving watering during the summer seasons of the first two years; no other irrigation was given. Under such resourcelimited conditions, the highest seed yield was about 500 kg per ha in the fifth year.

Regular irrigation was introduced in the next year and this resulted in luxuriant vegetative growth, followed by a longer flowering period and increased number of fruit pickings. The seed yield in the sixth year, when all the pickings are completed, is expected to be about 1.2 tons per ha per year. This yield is achieved under less than average conditions. Therefore, a seed yield of 2.0 tons per ha appears to be an achievable target with average levels of fertilisation, irrigation and management.

Intensive Jatropha production on good agricultural land with drip irrigation, high levels of fertilisers or manures and adoption of management practices such as regular pruning may yield 5.0 tons or more. However, this target has to be treated rather cautiously because plantations of this intensive kind have been established only within the past two years in India and they have not yet come into seed production. The seed yield range for Jatropha, quoted above, is from 0.5 tons to 10 tons per ha.

A farmer getting into Jatropha production cannot start off with an expectation of 5.0 or more tons. At the same time, a very conservative yield of less than 1.0 ton does not hold out any promise for a prospective grower to enter the field. A modest target that has a high probability of attainment is 2.0 tons per ha. The price at which Jatropha seed can be sold for oil extraction is generally Rs. 5/- per kg. Thus, a safe gross income target is Rs. 10,000 per ha. Setting this as the income potential, the establishment cost has to be kept to less than Rs. 7000/- and that on aftercare to Rs. 2000/- annually. The break even in such a situation would be in about six years and thereafter a net income of Rs. 8000/- per year can be expected.

Production Systems

Farm land: The economics of Jatropha cultivation, described above, may not appear appealing enough to attract would-be growers in large numbers. This initial hunch, however, is likely to get less intense on closer examination of facts. The actual income from vast areas of wasteland in the country is lower than what is possible with Jatropha cultivation. Most of these lands belong to small marginal farmers in dry areas who routinely take up cultivation of some crop during the rainy season. Jatropha is a suitable option for them. There are others who raise crops on a part of their land while the other part, usually the less fertile one, remains barren. Such land can be brought under Jatropha as it requires very little management after establishment and will not be a constraint on farmer's time and resources. It is also possible to enhance the role of Jatropha in existing production systems. A single row of Jatropha as live fencing is a common practice. This can be increased to 2-3 rows in a strip of 3.0 m along the farm boundary.

In a farm of 1.0 ha, this would amount to devoting 0.10 ha (0.25 acre) for Jatropha. Besides the service function of protecting the farm, this strip of Jatropha will produce about 200 kg of seed with very little additional resource commitment. Studies conducted at BAIF to examine the suitability of Jatropha for agroforestry purposes have also yielded promising results. When it was grown as the upper storey species, yield reduction on the annual crops horse gram and finger millet was negligible during the first four years. In fact, the annual crops seemed to benefit from the sheltering effect of Jatropha when there were strong dry winds. Being a small tree, Jatropha may also fit in as the under storey species with taller trees if the shade is not too heavy.

Common land: Large-scale Jatropha seed production on common wastelands has been highlighted extensively in popular articles. There have been limited attempts by the Indian Railways to raise Jatropha on land owned by it. Leasing of government land for the specific purpose of non-edible oilseed production is another suggestion put forward. The fertility status of common wastelands is such that intensive production with high levels of inputs may not be a viable option. At least until the soil fertility is improved, the cultivation approach has to be extensive on larger areas with modest levels of inputs. In such a scenario, the aim should be maximum returns and not maximum yields. This will require a systematic approach to establish and manage the large plantations. The target seed yield in extensively managed plantations on wastelands should be about 1.0 ton per ha.

REFERENCES

Dar, William D. (6 December 2007). "Research needed to cut risks to biofuel farmers" (in English), *Science and Development Network*. Retrieved on 2007-12-26.

Pae, Peter. "Jatropha plant's oil studied as jet fuel - Los Angeles Times". www.latimes.com. Retrieved on 2008-06-05.

6

Oli Palm Cultivation

The oil palms (Elaeis) comprise two species of the Arecaceae, or palm family. They are used in commercial agriculture in the production of palm oil. The African Oil Palm Elaeis guineensis is native to west Africa, occurring between Angola and Gambia, while the American Oil Palm Elaeis oleifera is native to tropical Central America and South America. The generic name is derived from the Greek for oil, elaion, while the species name refers to its country of origin.

Mature trees are single-stemmed, and grow to 20 m tall. The leaves are pinnate, and reach between 3-5 m long. A young tree produces about 30 leaves a year. Established trees over 10 years produce about 20 leaves a year. The flowers are produced in dense clusters; each individual flower is small, with three sepals and three petals. The fruit takes five to six months to mature from pollination to maturity; it comprises an oily, fleshy outer layer (the pericarp), with a single seed (kernel), also rich in oil. Unlike other relatives, the oil palm does not produce offshoots; propagation is by sowing the seeds.

Oil palms are grown for their clusters of fruit, which can weigh 40-50 kg. Upon harvest, the drupe, pericarp and seeds are used for production of soap and edible vegetable oil; different grades of oil quality are obtained from the pericarp and the kernel, with the pericarp oil used mainly for cooking oil, and the kernel oil used in processed foods. For each hectare of oil palm, which is harvested year-round, the annual production averages 10 tonnes of fruit, which yields 3,000 kg of pericarp oil, and 750 kg of seed kernels, which yield 250 kg of high quality palm kernel oil as well as 500 kg of kernel meal. The meal is used to feed livestock. Some varieties have even higher productivities which has led to their consideration for producing the vegetable oil needed for biodiesel.

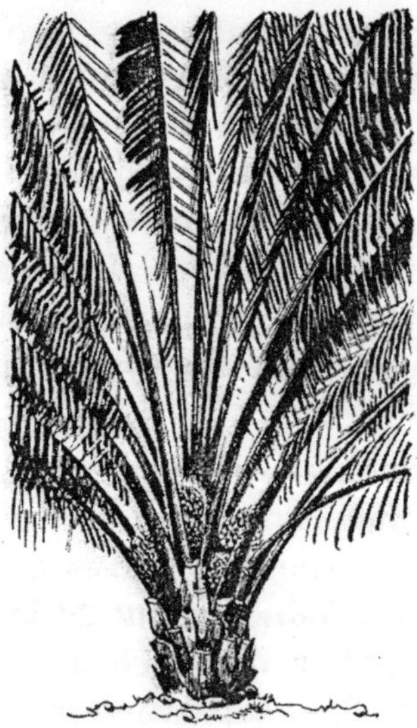

Oil plam tree

The African Oil Palm was introduced to Sumatra and the Malaya area in the early 1900s; many of the largest plantations of oil palms are now in this area, with Malaysia growing over 20,000 square kilometres. Malaysia claims that in 1995 it was the world's largest producer with 51% of world production. Palm oil and its fractions are practical and attractive choice for importers and food manufacturers, especially in 3rd world countries due to its price competitiveness, year-round supply, diversity and versatility for edible and non-edible applications.

The oil palm is a tropical palm tree. There are two species of oil palm, the better known one is the one originating from Guinea, Africa and was first illustrated by Nicholaas Jacquin in 1763, hence its name, Elaeis guineensis Jacq.

The fruit is reddish about the size of a large plum and grows in large bunches. A bunch of fruits can weigh between 10 to 40 kilograms. Each fruit contains a single seed (the palm kernel) surrounded by a soft oily pulp. Oil is extracted from both the pulp of the fruit (palm oil, an edible oil) and the kernel (palm kernel oil, used mainly for soap manufacture).

For every 100 kilograms of fruit bunches, typically 22 kilograms of palm oil and 1.6 kilograms of palm kernel oil can be extracted. The high productivity of the oil palm at producing oil (as high as 7,250 liters per hectare per year) has made it the prime source of vegetable oil for many tropical countries. It is also likely to be used for producing the necessary vegetable oil for biodiesel, an example being a planned refinery in Darwin, Australia which will import the palm oil from Indonesia and Malaysia.

The oil palm originated in West Africa but has since been planted successfully in tropical regions within 20 degrees of the equator. There is evidence of palm oil use in Ancient Egypt.

The world's largest producer and exporter of palm oil today is Malaysia, producing about 47% of the world's supply of palm oil. Indonesia is the second largest world producer of palm oil producing approximately 36% of world palm oil volume. According to US Departement of Agriculture, since 2005 Indonesia became the world's largest producer of Crude Palm Oil (CPO. Both nations are expanding their palm oil production capacity and the market continues to grow.

Worldwide palm oil production during the 2005-2006 growing season was 39.8 million metric tons, of which 4.3 million tons was in the form of palm kernel oil. It is thus by far the most widely-produced tropical oil, and constitutes thirty percent of total edible oil production worldwide..

The social and environmental impacts of oil palm cultivation is a highly controversial topic. There are multiple sources highlighting the positive and negative aspects of this industry.Oil palm is a valuable economic crop and provides a major source of employment. It allows many small landholders to participate in the cash economy and also often results in the upgrade of the infrastructure (schools, roads, telecommunications) within that area. However, there are cases where native customary lands have been appropriated by oil palm plantations without any form of consultation or compensation, leading to social conflict between the plantations and local residents. In some cases oil palm plantations are dependent on imported labour or illegal immigrants, and there are some concerns about the employment conditions and social impacts of these practices.

Biodiversity loss and the potential extinction of charismatic species is one of the most controversial issues in oil palm cultivation. The impacts of oil palm plantations on the environment is dependent on multiple factors, including the existence and compliance to environmental legislation, the pre-establishment habitat and corporate responsibility. In some states where oil palm is established there had been little enforcement of environmental legislation leading to encroachment of plantations into protected areas, encroachment into riparian strips, open burning of plantation wastes and release of palm mill pollutants in the environment. Some of these states have recognised the need for increased environmental protection and this is resulting in more environmental friendly practices.

Demand for palm oil has increased in recent years due to its use as a biofuel, but recognition that this increases the environmental impact of cultivation as well as causing a food vs fuel issue has forced some developed nations to reconsider their policies on biofuel to improve standards and ensure sustainability.

MODERN OIL PALM PLANTATION

A modern oil palm plantation needs a grower who has learned how to cultivate oil palms. Growing selected oil palms is not just a matter of picking the fruit; it is a modern crop. The grower must learn how to do his work well. The grower should ask for advice, so that he learns to do better and better. The grower must think about his work and plan it, so that he can always do his work at the right time. Selected oil palms give the grower much more work than the natural trees, but they yield much more. An oil palm begins to produce 3 or 4 years after it has been planted. During that time the grower must spend money and work hard, without harvesting any fruit or earning any money. Unless you apply fertilizers to the oil palms when they are still young, they will not grow well and you will have to wait a longer time before you can begin to harvest.You may also have to pay workers to help you look after the young plantation. Weeds must not be allowed to get in the way of the oil palms, and the trees must be protected from damage by rats and agoutis. All this work takes a lot of time, and this means that you may not have enough time to look after large fields of food crops.

Before you start an oil palm plantation, you must calculate carefully whether you will be able to pay all these expenses.To grow oil palms takes a lot of work. Before planting your oil palms, you have to clear the forest and remove the tree stumps. Then it takes a lot of time to sow the cover crop, dig holes in the plantation, take the seedlings out of the nursery, carry them to the plantation and plant them. All this work needs to be carefully done; you must take your time. Never hurry if you want to be successful with your plantation. Once the oil palms are planted, you must put wire netting around the young trees, you must spread fertilizer and keep watch over the plantation. Young oil palms need a lot of care. It is better to make a smaller plantation, but look after it carefully.

Once the oil palms have begun to produce, the fruit must be harvested at the right moment. If you cannot pick the fruit at the right moment, it becomes too ripe, many clusters will drop and the quality of the fruit will be less good. To grow oil palms takes much time and much care. Do not try to cultivate too large an area, or the work will be badly done. It is better to cultivate a small plantation and to do the work well. A small plantation that is well looked after can yield more than a large plantation that is badly looked after.

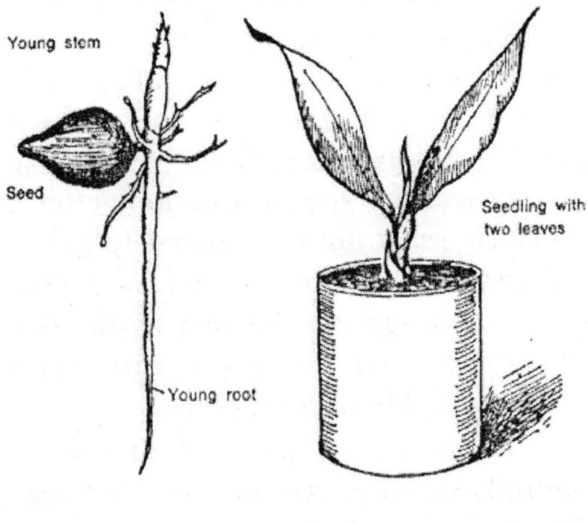

The oil palm may have a very long life. It is important to know about the life of the oil palm. If you know all about the life of the oil palm, you will understand better how to cultivate them. If you do not take care of the seeds, they will germinate only after several years. At the research stations, the seeds are kept in a room where it is always very hot. This makes the seeds germinate sooner, after 90 to 100 days.

Each seed germinated is planted in a small plastic container. A new leaf grows every month. The young seedling stays in the container for 4 to 5 months.When you see a left with two points (bifid leaf) coming up, transplant the seedling out into the nursery. The seedling stays in the nursery for 1 year. When it has about 15 green leaves, it is planted in the palm grove.The seedling is therefore 16 to 18 months old when it is ready to be planted in the palm grove.

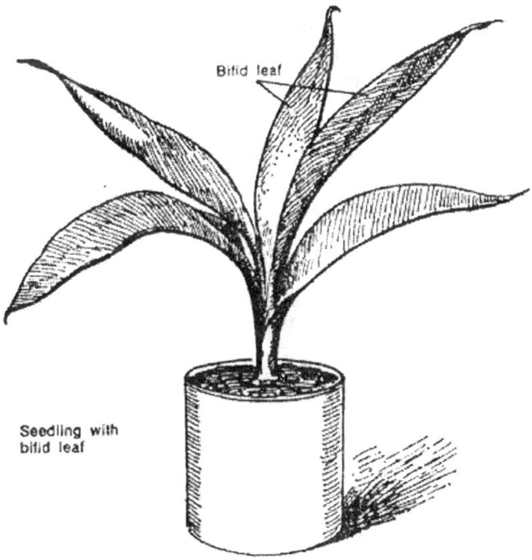

When the young oil palm has been planted in the palm grove, it produces male flowers.The flowers form at the base of each leaf. For several months, the oil palm produces only male flowers.After that, for several months, it produces only female flowers. The male flowers are grouped in spikes. The female flowers form other spikes. The male flowers fertilize the female flowers. Fertilized female flowers turn into a cluster of fruit.

Spikes of male flowers

The oil palm has no branches. It has a trunk and leaves. The trunk, sometimes called a stipe, is the stem of the palm. The growing point of the adult oil palm produces 20 to 25 leaves every year.

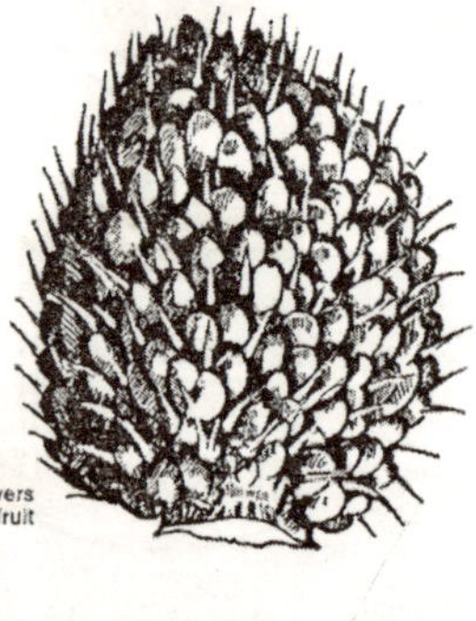

Fertilized female flowers turn into a cluster of fruit

It is most important that the growing point should produce many leaves, because there will be a flower at the base of each leaf. If there are many leaves, there will be many flowers. And if there are many flowers, there will be many clusters of fruit. The oil palm grows well and produces a lot in regions where it is very hot, where the sun is very strong, and where it rains a great deal.

Before getting the oil out of the fruit, the fruit must be separated from the spikelets. The fruits of the oil palm consist of the following parts:

— Pulp: the pulp is yellow; when the pulp is crushed it yields palm oil.
— Seed: inside the shell of the seed is the kernel; when the kernel is crushed, it yields palm kernel oil.
— The kernel also contains the germ.

The fruits of all oil palms are not the same.

— They are not all of the same size.
— The pulp is not equally thick in all of them.
— The shell is not equally thick.
— Some kernels have no shell at all.

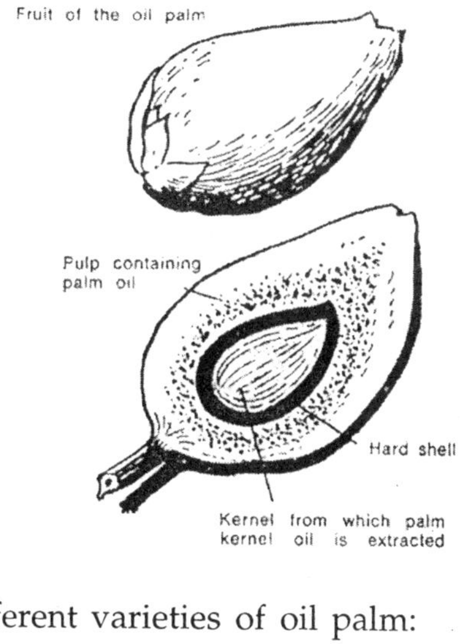

There are different varieties of oil palm:

— dura palms have kernels with a thick shell;
— pisifera palms have kernels with no shell;
— tenera palms have kernels with a thin shell.

When oil palms bear many and large fruit clusters, they yield a lot of oil. But to get a lot of oil, each fruit must also contain a lot of pulp, a shell that is not very thick, and a big kernel. Research stations have developed varieties of oil palm which produce many large clusters with fruits that have a lot of pulp, a thin shell and a big kernel.

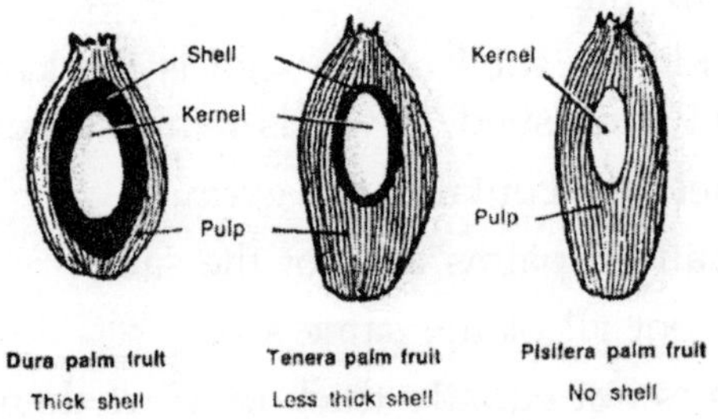

In order to get a lot of oil, the female flowers of a dura palm are fertilized with the pollen from a pisifera palm. Once they are fertilized, the female flowers turn into fruits. These fruits are of the tenera variety. The fruits of the tenera palm have a lot of pulp, a thin shell and a big kernel.

In traditional farming, nobody cultivates oil palms. People simply pick the clusters of fruit from the oil palms that grow in the forest. But these oil palms produce little. The oil is extracted by traditional methods, and a lot of oil is left in the pulp and the kernel.

But nowadays oil palms are grown on modem plantations. These contain selected oil palms with big yields. The clusters of fruit are sold to mills which extract all the oil from the pulp and the kernels. These oil palm plantations bring in money:

— for the growers who sell the fruit,
— for the workers who work in the mills,
— for the government which can sell the oil to foreign countries.

The growers can also earn money by raising beef cattle. Beef cattle can be fed with the green fodder from the cover crop grown in the palm groves.The grower can also feed his cattle with palm-kernel oil cake, that is, what is left over after extraction of the palm kernel oil. Oil palms are cultivated in the regions where they grow well and where there are oil mills. The oil palm grows well where it is hot all the year round: between 25 and 28 degrees C. If the temperature drops, the oil palm produces fewer leaves and is more often attacked by diseases. It therefore yields less. A hot temperature enables the oil palm to make many leaves and to produce many clusters of fruit.

Where there is a lot of sunshine, there will be strong photosynthesis, provided the oil palm is in soil which gives it water and mineral salts. The leaves grow large, the fruit ripens well, and there is more oil in the fruits.

If it does not rain much, or if it does not rain for several months, the leaves do not grow well. If there are few new leaves, there are few flowers and few clusters of fruit. Where the soil is flat, deep, permeable and rich. The oil palm needs a flat soil. If the soil is not flat, transport is difficult and costs a lot. Erosion is severe; the water carries away the earth. The oil palm needs a deep soil.

The roots of the oil palm cannot develop il they meet a hard layer. They cannot take up water and mineral salts that are deep down. If the oil palm does not have enough water, yields are low. The oil palm needs a permeable soil.

The oil palm does not grow well if water remains around its roots for too long. The oil palm needs a rich soil. In order to produce many large clusters of fruit, the oil palm needs a lot of mineral salts. If the soil is poor, mineral salts can be added by applying fertilizers.

With traditional methods, a lot of oil is left in the pulp and the kernels. The machines of the oil mills extract all the oil contained in the pulp and the kernels. Selected oil palms produce many clusters of fruit. To get all the oil out of these clusters yourself, you would have to spend a lot of time. Before planting selected oil palms, make sure you can sell the fruit clusters to a mill.

Where business companies or extension services can give the grower advice. It takes much money and work to make an oil palm plantation. The grower must use modern methods in order to pay for his expenses and earn money. He will need advice on:

— how to choose the site for his plantation
— how densely to plant it
— how to look after the plantation
— how to apply fertilizers
— how to protect the oil palms against disease.

THE OIL PALM NURSERY

It is difficult for a grower to make the seeds of oil palms germinate. Growers buy young seedlings which already have four or five leaves. Seedlings can be bought from research stations or extension services. The young seedlings are then put into a nursery.

The nursery is a small plot in which the young oil palms develop. When the oil palms are big enough, they are planted out in the palm grove. Nurseries cause a lot of expense and need much care. It would be very expensive for one grower alone to make a nursery; it is better to make the nursery jointly with other growers. It is very important to make a success of the nursery, so as to get fine young plants.

A seedling that has not grown well in the nursery will make a poor oil palm. To have fine seedlings in the nursery you must:

— choose a good site and prepare it well,
— choose the finest seedlings,

— water them, protect them against erosion and weeds, give them fertilizers, protect them against insects and diseases.

How to Make a Nursery

Choosing the site

The soil should be fairly rich and well prepared. It is best to clear a bit of forest for the nursery plot. If you clear a forest site for the nursery, pull out all the trees and burn them. Burning all the wood helps to control certain diseases which might attack the roots of the oil palms, and it also makes the soil more fertile. Spread the ashes all over the plot.If you put the nursery on a field which is already cultivated, pull up all the old crops: cocoa trees, coffee trees, oil palms.

When the site is well cleared, it needs deep tilling. You should till 40 centimetres deep with a hoe or a tractor. To improve the soil structure, you can then sow a green-manure crop, like Centrosema or Crotalaria. When these crops have grown, work them into the soil by tilling again. Then apply fertilizers: 500 kilogrammes of dicalcium phosphate per hectare.

Layout of Paths and Nursery bed

Nursery bed is the name for the strip of soil where the oil palm seedlings are planted.It is best to make the nursery on flat ground. But, if the ground slopes, the beds must lie across the slope.

The beds should be 45 metres long and 3.5 metres wide. The soil of the beds should be well worked to make it quite flat. After that, apply a dressing of fertilizer. For instance, at La Mé, Ivory Coast, 250 kilogrammes of 10:10:20 fertilizer are applied per hectare.

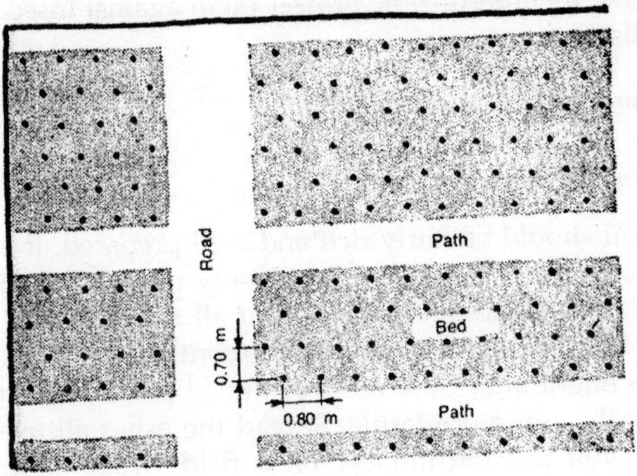

Making holes for seedlings and transplanting. To know where to make the holes for your seedlings, make a pattern. At the places where you have put your little pegs, make a hole with a Richard plant setter. Then put a seedling with its ball of earth into each hole.

You must give the seedlings a lot of water. But do not water when it is hot; it is best to water in the evening and the morning. To protect the soil against erosion, mulch it.

Cover the ground with herbage or cluster residues. Leave a ring of 20 centimetres of unmulched ground around each seedling. If you mulch with cluster residues, put them down only three months after transplanting, so that the insects do not attack the young leaves. If you mulch with herbage, you must replace the herbage when it rots. Then hoe the soil. If you cannot get enough water for the seedlings, transplant them into the nursery at the beginning of the rainy season. At the end of the rainy season, the seedlings will be strong enough to get through the drought.

Pattern for nursery planting

In certain regions shelter has to be put up over the nursery. This protects the young seedlings from a disease called blast. These shelters are made with posts and bamboo sticks. To make the shelters more solid, put two posts together. The posts should be 2.5 metres high. The bamboo sticks are tied to the posts with lianas. Finally, put palm fronds over the bamboo sticks.

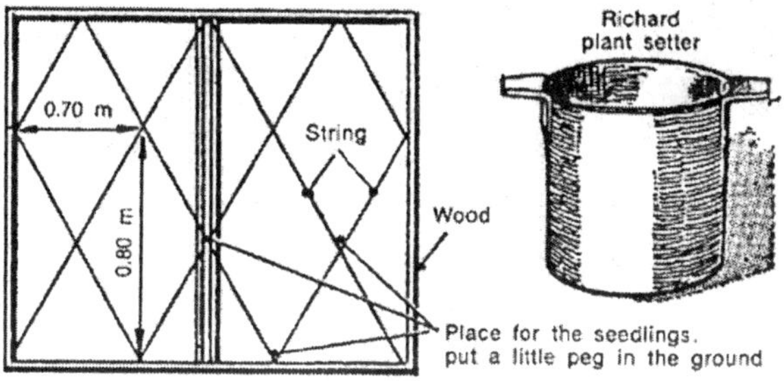

In Benin, shade for the young oil palms is provided by planting castor-oil plants in the nursery. If you make a shelter, you need not mulch, but you must hoe very often. Get rid of all the weeds, and always keep the soil loose. Three months after transplanting, if the seedlings have grown well, apply monthly to each plant 15 to 20 grammes of a mixture of ammonium sulfate and potassium chloride.

Spread the fertilizer mixture in a ring 10 centimetres from the seedling and water. Hoe to work the fertilizer into the ground. The seedlings stay in the nursery for about 1 year. You must plan, therefore, to make the nursery about 1 year before you want to plant your palm grove.

PREPARING THE SITE

In big plantations the forest trees are pulled up with the help of big machines. But often the grower cannot use these machines. Begin by marking out the boundaries of the plantation. Next, cut down as many trees as you can around the plantation and take all the trees away, so that the fire cannot burn down the whole forest. This is called making a firebreak. Then set fire to the plantation site.

When the fire is over, the soil is bare. You must protect the soil against the sun, for the sun burns the soil and destroys the humus. At the beginning of the rainy season, sow a cover crop;

— a mixture of Pueraria javanica
— Centrosema pubescens, and
— Calopogonium muconoïdes.
— Sow 4 to 6 kilogrammes of seed per hectare.

Pegging the Planting Pattern

When the soil of the plantation has been well cleared by fire, peg out the places where you will plant your seedlings. To be sure to plant at the right density, you must peg out carefully before planting. Then you will be sure of having always the same distance between rows and in each row the same distance between oil palms.

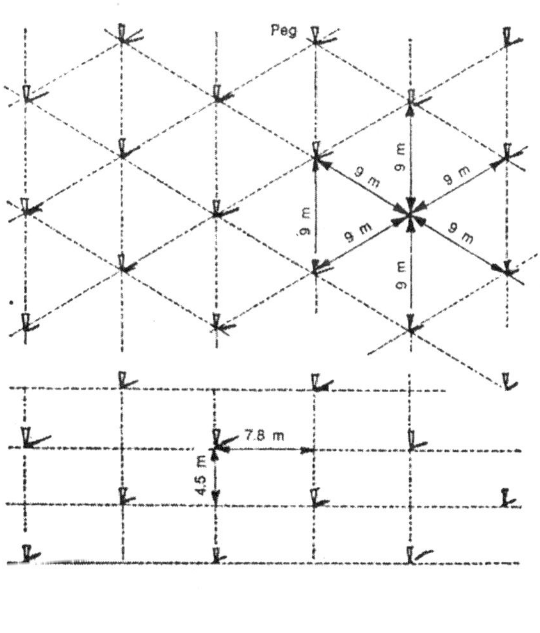

Planting pattern for a plantation

To get a good yield, you must plant the oil palms at the right density. If the oil palms are planted too close together, the roots get in each other's way, and the leaves do not have enough air and sun: the yield will be low. If the oil palms are not planted close enough together, each separate tree produces much, but the roots do not use all the soil: the yield per hectare will be low.How to peg out the planting pattern.

Trace lines across the slope and put in your pegs in straight lines; leave 7.8 meters between rows and 9 metres between pegs. In this way you can plant 143 oil palms per hectare; this is the best density. Pull out tree stumps and remove fallen trees close to the pegs, because these stumps and trees would interfere with the oil palms.

One month before planting, dig a hole at each place where you have put a peg. The hole should be 0.60 metre long and wide, and 0.80 metre deep. When you are digging the holes, cut any roots that you find in the soil. Do not mix the soil from above and the soil from below.

Soil from the top in one heap
Soil from the bottom in another heap

A few days later fill in the holes with the earth you have dug out. At the bottom of the hole, put the soil you have dug out from the top, and at the top put the soil you have dug out from below. Fill the hole well, so that no saucer shape forms on top.

The right time to plant is the beginning of the rainy season. In this way the young plants can develop their root system before the dry season arrives.

Choose the biggest and the best-grown seedlings. Leave in the nursery any seedlings that are small or badly grown.

Cut away all the dry leaves and the tip of leaves that are too long. Put grease over the cut ends where you have removed leaf tips.

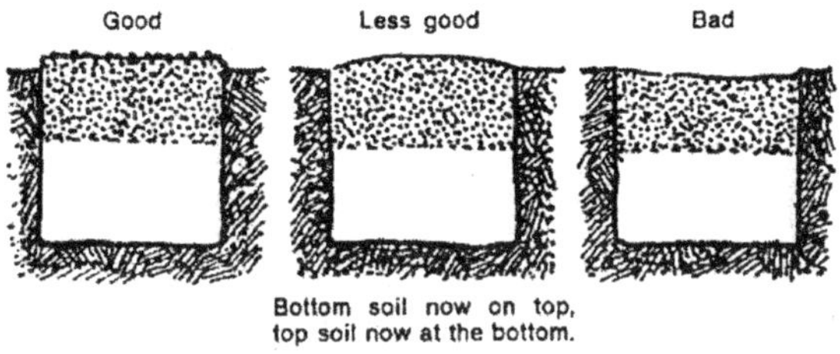

Do not lift the seedlings long before you plant them. Lift and plant them in the course of a single day. In order to keep a big ball of earth around the roots, lift and plant the palm oil seedlings with a plant setter. If you use a Socfin plant setter, place the seedlings after lifting into a wooden box in order to carry them to the plantation.

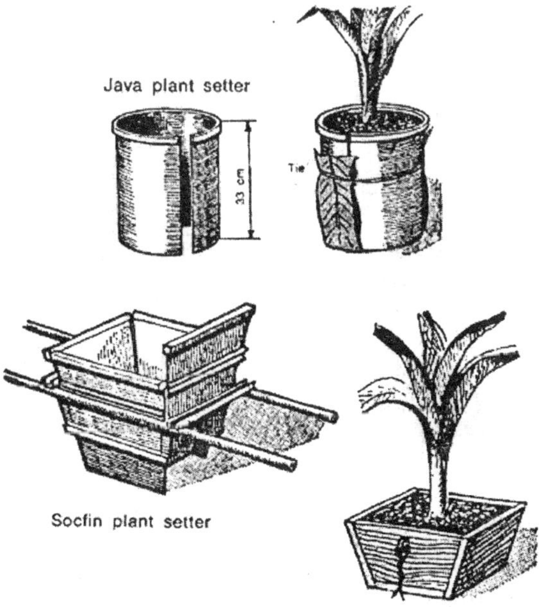

If you use a Java plant setter, leave the seedling inside the plant setter when you carry it to the plantation. The seedling is tied into the plant setter. Make the holes for planting in the plantation, with the plant setter you use for lifting the seedlings from the nursery beds. Do not make the holes several days before planting. If you make the holes too long before planting, the rain may wash earth from the sides into the hole, or the sun may dry out the earth on the sides.

The ball of earth around the roots must be level with the soil of the plantation. The earth must not form a hollow around the crown. Fill in with earth the space between the sides of the hole and the root ball. Remove with a little stick all the earth that has fallen on the leaves.

Putting Wire Netting

Certain animals may eat the young oil palm seedlings. To protect the seedlings, surround them with wire netting. Leave the wire netting in place for about 18 months.

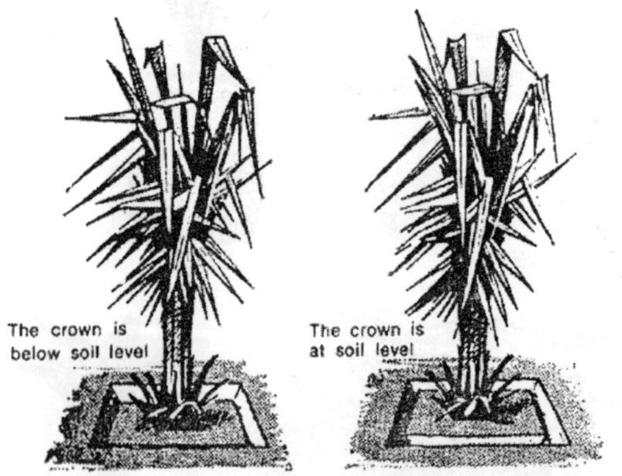

When you have put your wire netting in place, spread a mulch 20 centimetres thick around the seedlings. This mulch prevents the soil from drying out, and prevents weeds from growing. Use dry herbage, and spread it 15 to 20 centimetres thick at a distance of 30 to 40 centimetres around the crown. A few months after planting, apply fertilizers. The right time to apply fertilizers is near the end of the rainy season. The recommended dose of fertilizers for each plant is: 250 to 500 grammes of ammonium sulfate and 250 grammes of potassium chloride.

Spread the fertilizers in a ring underneath the longest leaves. After you have spread the fertilizer, cover it with a little earth. If there is a mulch around the seedlings, remove the mulch before applying the fertilizer. Afterwards spread a fresh mulch of dry herbage.

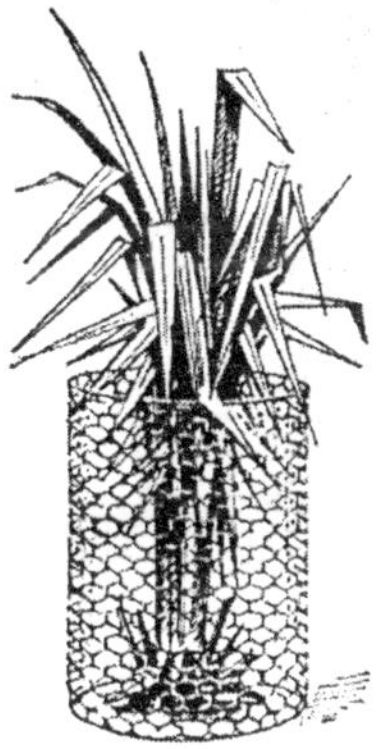

Wire netting in place

LOOKING AFTER THE PLANTATION

You must remove the weeds around the young oil palms. This work is done with a hoe or a machete. During the first year, cultivate 6 times. Remove all the weeds for 2 metres around each stem. During the first months, the weeds between rows have to be cut. If you leave the weeds, the cover crop will not grow well.

You must always cut away the dry leaves of the oil palm. In order to cut the leaves without damage to the oil palm, your tools must be well sharpened. Cut the leaves very close to the stem, so that no other plants can grow in the axil of the cut-off leaves. Remove from the trunk any plants (ferns) that may grow in the axils of the leaves. Remove also the male flowers.

The oil palm needs a lot of mineral salts to form its leaves and fruit clusters. When the oil palm is young, it needs above all nitrogen. When the oil palm has begun to produce, it needs a lot of potash. Potash increases the number of fruit clusters, and makes them bigger.

PROTECT AGAINST INSECTS

To protect the young trees, put in the axil of the leaves a mixture of sawdust and BHC. To avoid dangerous attacks, be very careful not to wound the trees. The insects may lay their eggs in the wounds of the oil palm. There are other insects, but it is difficult for the grower to control them.

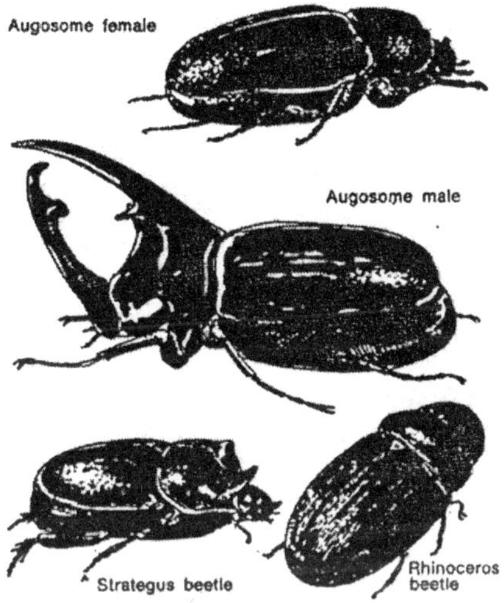

Oil palms may also be attacked by rats and agoutis. Rats and agoutis can eat young oil palms.

Protect your young oil palms against agoutis by wire netting round each tree.

As a protection against rats, you can place little bags with poisoned maize near the oil palms.

Oil palms may be attacked by several diseases. If you see distorted leaves or leaves that break, ask for advice from the extension service.

HARVESTING

Harvesting needs much time and much care, because only those fruit clusters which are cut at the right moment yield a lot of good-quality oil. You must go through the plantation many times to pick the ripe clusters.

A cluster is ripe for harvesting when the fruits begin to turn red, and when 5 or 6 fruits drop to the ground. If you wait too long before harvesting the clusters, harvesting takes much more time, because you must pick up all the fruits that have dropped to the ground. The fruits will also yield less oil, and the oil will be of less good quality. If you do not wait long enough before harvesting the clusters, the fruit will not be ripe enough. It will be more difficult to separate the fruits from the clusters and the clusters will yield less oil. The clusters can be cut off with different tools.

— *For oil palms 4 to 7 years old.*

Cut the clusters with a chisel.

Slip the chisel between the stem and the leaf; in this way you can cut off the cluster without cutting the leaf below it.

— *For oil palms 7 to 12 years old*

Cut the clusters with a machete.

If the clusters are too high up, climb up the tree by putting your feet on the base of the leaves.

— *For oil palms older than 12 years*

Cut the clusters with a long-armed sickle.

If the clusters are too high up to be cut with the longarmed sickle, use bamboo ladders, or else climb up the tree with a belt; you can also wear spiked shoes.

Any clusters that have dropped to the ground should be collected in a basket. Fruits that have come loose must also be picked up.

REFERENCES

Amiruddin, Mohd, N., "Palm oil products exports, prices and export duties: Malaysia and Indonesia compared," *Oil Palm Industry Economic Journal,* Vol. 3 No. 2: 21-31, 2003.

Basiron, Y., N. Balu, and D. Changrmohan., "Palm oil: the driving force of world oils and fats economy," *Oil Palm Industry Economic Journal,* Vol. 4 No. 1: 1-10, 2004.

7

Production of Mustard-Rapeseed

Mustard-Rapeseed is a group of oilseed crops which assumes the significance in Indian national economy by occupying the second position next to groundnut and is considered as a 'cash crop'. Biologically, the rapeseed and mustard plants belongs to the family Crucifarae and under the genus *Brassica* with large number of species and sub species cultivated in India. Being a number of *Brasica* genus, the rapeseed is closely related with mustard.

The word "rape" comes from the latin word "rapum", means 'turnip'. On the other hand, the word 'mustard' is derived from latin word "mustum" or "must", which denotes 'expressed juice of grapes and ardens" means "hot and burning". The mustard-rapeseed is a versatile group of plants used in various ways historically. The mustard and rapeseed are closely related each other and can share the same growing areas.

The cultivation of mustard-rapeseed dates back to 2000 B.C. both in sub-tropical and tropical countries. The mustard- rapeseed plants grow all over the world but their cultivation is mainly confined to India, China, Canada, Germany, France, Australia, USA, etc. The rapeseed was cultivated in Europe since 13th century. *Brassica* group of plants yields the oils which characteristically have large amount of long chain fatty acid called as "erucic acid".

Prior to 1960s, the erucic acid content of this oil was not taken as important factor for evaluation of its oil quality.

However, from 1960s, considerable research have been done to develop low erucic acid varieties as the researchers found that consumption of erucic acid in oil has negative effect on health. Due to this, new varieties with low erucic acid content were developed in Canada in late 1960s and early 1970s. During 1978, the rapeseed industry in Canada identified the new rapeseed varieties with low erucicacid content and glucosinolates as 'canola' quality standard. Production of canola is largely supplied by two species ie., *Brassica napus*. L and *Brassica rapa* L. and to alesser extent by *Brassica juncea* Coss. (Brown mustard) and *Sinapsis alba* L. (YellowMustard).

World Production of mustard-rapeseed has increased faster over past two decades than any other oilseed. At present, canola/rapeseed has now become the second largest produced oilseed in the world after soybean. In India, the mustard - rapeseed is the most important oil seed crop after groundnut accounting around 25 per cent of total oilseed production. It is one of the important oilseed crop of the Indo -gangetic plains. Indian mustard (Rai)cultivation has occupied about 85-90 per cent of total area under cultivation of mustard - rapeseed.

The traditional mustard-rapeseed grown in India contains high amount of erucic acid and glucosinolates and this does not conform the international standard 'canola' quality. Hence, at present the government agencies in Punjab are promoting the cultivation of hybrid mustard rapeseed namely, 'Hyola' as a drive for the crop div ersification programme. Punjab Agro initiated the contract farming of hyola in around 10,000 acres of Punjab during 2002-2003.

Mustard-Rapeseed

The oil obtained from mustard seeds is known for it's

culinary fats for over 3000 years by the Indians, West Asians (Indian sub-continent) and Chinese. Being a rich source of unsaturated fatty acid and with low concentration of saturated fatty acid, now-a-days the oil of mustard- rapeseed has become nutritionally better than other oilseeds specially after the introduction of 'canola' quality mustard-rapeseed in North America, European Union and Australia.

The term 'canola' is a registered trademark of 'Canada Canola Association' and refers to the mustard-rapeseed having low erucic acid and glucosinolate, which has gained a tremendous acceptance worldwide. Canola oil is widely used as cooking oil, salad oil and for preparation of margarine in different parts of the world. World trade of canola quality rapeseed has achieved a tremendous growth. Globally, the volume of canola/rapeseed export is the second largest volume of oilseed traded following soybean.

In India, the oil obtained from mustard- rapeseed accounts for 2/3rd edible oil consumption in the country. In India, the oil is traded in kacchi-ghani type for it's traditional characteristic flavour and to some extent in refined form. The projected demand for oilseeds in India by 2020 is around 34 million tonnes which is to be met by mustard-rapeseed. It is estimated that about 90 per cent of domestic production of mustard-rapeseed is crushed for extracting edible oil, which is mostly traded and consumed in northern, north -eastern eastern and central India.

The recently promoted 'canola' quality hybrid rapeseed namely 'Hyola' has a good potentiality as a profitable enterprise for farmers as it gives a higher yield, more oil content, healthy export quality oil and an assured market. Besides, the utilities of oil obtained from mustard-rapeseed, the seeds, sprouts, leaves, tender plants are also useful to human health, when they are consumed as spices and vegetables. They contain selenium, calcium, magnesium, iron, phosphorus, zinc, magnesium, manganese, etc.

PRODUCTION TRENDS

Mustard-rapeseed is grown in more than 50 countries in Asia, Europe, America and Australia with a production at about 36778 thousands tones during 2003. Out of 53 mustard - rapeseed growing countries in the world, India and China together accounted for 42 percent of total production. However, out of global production, Canada, China, Germany, India, France and Australia share more than 80 per cent of total production among which India contributed 10 per cent to the total world production as can be seen from following chart. The country wise production during the period 2001 – 2003 have been presented in Table 1.

Table I. Production of major mustard- rape seed producing countries during 2001 to 2003

(Production in thousand tonnes.)

Country	*Production*		
	2001	*2002*	*2003*
China	11345	10565	11425
Canada	5122	4332	6895
India	4187	5083	3842
Germany	4164	3853	3642
France	2879	3319	3343
Australia	1757	841	1622
Others	6837	6539	6009
All World	36291	34532	36778

In India, among all types of mustard - rapeseed, Indian mustard is cultivated in Assam, Gujarat, Haryana, H.P., M.P., Orissa, Punjab, Rajasthan and West Bengal. It's cultivation has been extended to southern states on a limited scale. The brown Sarson is grown in Kashmir and Himachal valley, whereas, the yellow sarson is grown in Eastern U.P., Assam, Bihar and West Bengal. Toria is a short duration crop cultivated largely in Assam, Bihar, Orissa and West Bengal.

Taramira grown in drier parts of North west India comprising of Haryana, Punjab, Rajasthan and U.P. However, the seven states i.e, Rajasthan, U.P, Haryana, West Bengal, M.P, Gujarat, Assam contribute a lion's share of production as can be seen from the Table 2.

Table 2. State-wise production of mustard-rape seed during 2000-2001 to 2002- 2003

State		*Production*	
	2000-2001	*2001-2002*	*2002-2003*
Rajasthan	1312.8	1943.0	1318.3
U.P.	945.7	845.4	759.1
Haryana	554.0	796.0	694.0
West Bengal	417.0	336.9	328.5
M .P.	323.6	459.2	210.3
Gujarat	230.6	292.1	172.3
Assam	141.2	137.0	129.8
Others	262.3	273.0	305.8
All India	4187.2	5082.6	3918.1

Qualities of Mustard Rapeseed

Canola quality

'Canola' is genetically altered and improved version of rapeseed. 'Canola' quality varieties are presently developed from either or two species of *Brassica nupus* and *Brassica campestris*. 'Canola' is a registered trade mark of Canadian Oil Association. The name 'canola' denotes the seed having less than 2 per cent erucic acid in it's oil and less than 30 micro -moles of glucosinolate per gram of it's deoiled meal. The rapeseed, which does not conform to the above qualities is not termed as "canola", which is a quality standard and not a biological classification.

Canola varieties, which are having both low erucic acid and glucosinolates are also termed as 'double low' or '00' rapeseed or sometimes as LEAR (Low Erucic Acid Rapeseed). The rapeseed oil was used in Europe and Asia as a cooking medium and for other uses since long, but this oil contained high amount of erucic acid and glucosinolate, which was not good for health.

But with development of canolacultivars in Canada, it's use has increased manifold. After 1970s, the 'canola' quality oil has gained the acceptance worldwide. Now-a-days, it is major edible oil in Australia, Japan, Canada, and many European countries.

Hyola

It is a "canola" quality hybrid rapeseed mustard recently introduced in India. The Hyola (variety PAC-401) is only hybrid 'canola' quality gobhisarson notified by Govt. of India after extensive trials by Indian Council of Agricultural Research (I.C.A.R.). It has the yield potential with high oil percentage. In current crop diversification programme, Hyola is gaining wide acceptance among the farmers in Punjab. The acreage of Hyola has already crossed 50,000 acres.

Besides farmers, it has also been accepted by the traders and consumers. As a result, Punjab Agro Industries Corporation Ltd (PAIC) is promoting the Hyola cultivation through contract farming system and has identified the Hyola (variety PAC 401) as a suitable alternative in place of wheat during rabi season. By realising it's benefits, the Govt. of Punjab has given more emphasis on it's cultivation and set a target to extend the cultivation to more than 1 lakh acre by 2004 -05 and which will be 5,00,000 acres by 2007-08.

Teri-Uttam

The scientists of 'The Energy and Resources Institute' (TERI), New Delhi have developed a new variety of gobhi-sarson and named as "Teri-Uttam". Oil obtained from Teri-Uttamis nutritious with better fatty acid profile and is comparable to the Internationally recognised "Canola " oil. This variety has shown immense potential because it contains high oil content, early maturity period with in 135 days and more than 20 per cent yield as compared to the *Brassica napas* National check variety GS-1. It has also potential of earning foreign exchange as it's oil meal contains low glucosinolate.

POST HARVESTING

Losses of Mustard Rapeseed

Losses of mustard - rapeseed occur during post harvest operations like handling, transportation and storage by producer's, trader's, wholesaler's level, which vary from 0.2 per cent to 2.0 per cent.

Different losses of mustard – rapeseed may occur as following:

— Loss in weight in dry season.

— Due to improper method of harvesting and ignoring the symptoms of harvesting, deterioration of the quality of seeds of mustard -rapeseed is occurred.

— During handling/ lifting of bags, use of too much hooks by labourers causing spillage loss.

— Due to rodent attacks on bags which results in spillage waste and losses during storage.

Harvesting Care

Maturity period of harvesting for mustard-rapeseed:

Sl. No.	Type of Mustard-Rapeseed	Period of maturity for harvesting
1.	Toria	70-100 days
2.	Indian mustard	105-160 days
3.	Yellow sarson	120-155 days
4.	Brown sarson	100-235 days
5.	Taramira	140-150 days

Following care is to be taken during harvesting.

— The harvesting of toria and rai should be done when the pods begin to turn yellow. The sarson is less liable to shattering hence can be left in the field for relatively longer period and can be harvested it in comparatively dry ripe stage.

— When harvesting is done manually, care is to be taken to harvest the entire plant by pulling out or by uprooting the plant a few cm. above the ground by sickles.

— Harvesting should be done in the early hours during morning as the moisture accumulated during night prevent the splitting of pods.

— Crop should be harvested when the pods have lost their greenness but not at the dead ripe stage as it may cause the shattering of pods.

— When grown as mixed crop with food grain, then both the crops should be harvested separately.

Following care is to be taken after harvesting:

— After harvesting to separate the seeds from plants, threshing is to be done always on pucca floor instead of kuccha floor because threshing on kuccha floor increases the amount of impurities of soil, stone, dust, dirt particles mixed with seeds.

— After threshing, the seeds separated should be subjected to winnowing in which the mixture of seeds and chaff is allowed to fall from a height by which the chaff is blown away and seeds are gathered on the floor.

— Pack the seeds of gathered on floor in a sound B-twill gunny bag.
— Before transportation, drying of harvested plant in sun is also advisable since drying in farmers field involves low cost whereas the same operation in the market involves not only higher cost but also the unavoidable transportation of moist seeds.
— Care should be taken as not to sent 'moist 'oilseeds to the market. Because, it affects the quality; oil content and crushing quality of the seeds. Hence, dryharvested plant in sun for four to ten days by tying up into small sheaves.

Grading

It is the sorting out of homogenous lots according to the laiddown quality specifications. Produce is graded according to fixed tolerance limits for various quality factors which are universally accepted.

Grading benefits the oilseed farmers / traders / consumers:

— It facilitates to get higher price for the produce being graded according to specific grade standards, which are well known to both the parties.
— It widens the marketing process because buying and selling can take place between two parties at distant places and reduce the cost of marketing and storage losses.
— It facilitates the keeping quality of the produce and easy finance, when the produce is stored and getting claims settled by the always or insurance organisations.

— It facilitates the future trading and help the consumers to get standard quality of produce at fair prices.

Quality factors for grading of mustard-rapeseed:

— Foreign matter, immature, shrivelled/dead seeds, admixture of other oilseeds, weevilled/damaged seeds, moisture content etc.

— It should be free from fermented, musty odour, insect/fungus infestations, any other toxic seeds, seeds of *Argemone maxicana*, rodent hair and excreta.

Steps for grading

— *Cleaning and sorting:* After arrival, entire produce should be cleaned and sorted manually /mechanically to separate the foreign matter, dust, dirt and stone particles, admixture of other seeds, husks, immature split, shrivelled, damaged and diseased grains.

— *Packing and sealing:* The cleaned and sorted homogenous produce should be filled in gunny bags and then packed and sealed.

— *Sampling and analysis:* For successful grading of a lot, the drawing of truly representative sample is essential and it is done in such a way so that it reveals the exact composition of the commodity. Subsequently, the grading process is done by physical analysis of the samples and in addition to it, is analysed separately byseparate instruments for determining oil content and moisture percentage of seeds.

Features of sampling

1. *Primary sample:* Each sample drawn from the heap or bag by parkhi or tubesampler from a single position of the lot.
2. *Composite sample:* Primary samples drawn from the same lot shall be thoroughly mixed and blended to

form homogeneous composite sample in a sample divider.

3. *Test sample:* One portion of composite sample weighing 500 gms is packed in cloth bag.
4. *Sample for Moisture:* Part of the composite sample weighing about 250gms. packed in polythene bag and heat sealed kept in airtight container.
5. *Labelling of sample:* Appropriate labels are attached with cloth bagged and polythene bagged samples showing the following particulars:
   a) Name of the commodity and variety
   b) Lot number
   c) Quantity, whether in bags or in bulk
   d) Place and date of sampling
   e) Details of wagon/truck/warehouse in the case of bulk samples
   f) Name of sampling officer
   g) Signature

Hygiene

Mustard and Rapeseed, after cleaning and sorting, and before further processing :

— Shall be free from microorganisms in amounts which may represent a hazard to health.

— Shall be free from parasites which may represent a hazard to health.

— Shall not contain any substance originating from microorganisms, including fungi, in amounts which may represent a hazard to health.

Adulterants and toxins

The Arg emone mexicana Linn is originally an American herbaceous annual which is well known through different names in different languages and belongs to family and characteristics as follows:

English – Mexican prickly poppy

Hindi – Shialkanta, Satyan ashi

Bengali – Shialkanta

Gujarat – Darudi

Tamil – Kutiyotti

Sanskrit – Srigalkanta

Family :- Papavaraceae

Argemone seeds are blackish in colour. Surface is wrinkled whereas mustard seeds_are similar in colour but having smooth round surface. The *Argemone* seeds yield 22-36 per cent non edible, nauseous, bitter, pale yellow oil, which is known as *Argemone* oil. It contains two alkaloids which are responsible for dropsy disease.

Symptoms of dropsy: Symptoms of dropsy are acute nausea, vomiting, loose motions, bloated stomach, involvement of kidney, swelling of hands and feet known as Oedema. In extreme cases, it is reported that glaucoma and death due to cardiac arrest occurs. The line of treatment is symptomatic and it has been suggested that diuretic, bio antioxidants, steroids, vitamins, calcium and protein rich diet have beneficial effect on epidemic dropsy cases.

Preventive measures

The mustard oil be consumed which has undergone thorough quality tests. Following types of laboratory tests are used for detection of *Argemone* oil:

Screening test - Nitric Acid

- Ferric Chloride test

Confirmatory test - Paper Chromatography test

Grading at producer's level

In order to ensure proper price to the producers as well as gaining confidence of the consumers the mustard - rapeseed should be graded at producers' level. Some of the varieties like Maghi-Lahia, RH-30, Pusa Bold fetch good price in market. In order to ensure the correctness of description of particular variety, the grading should be done at producer's level. Grading not only facilitates the dissemination of market information but also facilitates the distribution at various stages.

The scheme of 'Grading at Producer's level' was introduced in 1962-63 by the Directorate of Marketing and Inspection (DMI). The main objective of the scheme is to subject the produce to sample tests and assign the grade before it is offered for sale. The programme is being implemented by state governments for which 1345 grading units have been established in India up to 31/3/2001.

Packaging of mustard - rapeseed

— *Plastic film bags:* Two types of polythene films are available viz. low-density polyethylene film (LDPE) and high density polyethylene film (HDPE) for packaging purposes. But in most cases, LDPE is used as plastic film bags for packaging mustard -rape seeds. It protects the oilseeds against dust, dirt and moisture. It is convenient for packing of smaller quantities of oilseeds like 1kg., 5kg., and 10kg. packs.

— *Corrugated fibre board :* These are paper board cartons used for keeping the plastic bags filled with mustard - rapeseed. It protects the oil seeds from dust, dirt and to some extent from moisture.

— *Jute bags:* Gunny bags made of jute are widely used by producers, traders, processors, etc,. for packing of mustard - rapeseed. Traders use gunny bags for storing oilseeds and keep them in go-down by stacking. Seeds of mustard – rapeseed are generally packed in 'B' twil jute bags of 95kg., capacity. Sometimes the oilseeds are stored in 50kg. jute bags also.

Criteria for selection of packaging material for mustard - rapeseed:

— It should be specific to the characteristics of the produce.
— It should be suitable according to transportation and storage method.
— It should be suitable according to climatic and environmental conditions.
— The material must provide protective strength to the produce.
— It should be safe to handle during transportation.
— The material should be economical, readily available, easy to handle and store.
— It should be convenient and suit the need of the customer.
— It should be attractive for display.
— It should be environment friendly and biodegradable.

The packages of mustard-rapeseed should be properly labelled with following marking:

— The name of the commodity, brand name, trade name of the produce.
— Name and address of the packer / farmer
— Net weight of the content in the package.
— Batch No. / lot No./ Code No.
— Date of packaging indicating the month and year
— Year of harvest, if available
— Grade of the material as per national grade standards
— Name of the producing country (to be mentioned in case of export).

For scientific packaging of mustard - rapeseed, following care should be taken:

- The produce should be packed in suitable packaging material as approved by the competent authority
- Quality of the packaging material should conform to the requirements as laid down under PFA standards as amended from time to time.
- The size of the pack shall be as per the provision under the standards of Weights and Measures Act.
- The oilseeds should be packed in clean, hygienic bags of any material, which does not affect the produce and prevents it from absorbing moisture.
- The packing material used should have sufficient aeration facilities.

Modes of Transportation

Pathways

Head load: Among all modes of transportation, the head load is the convenient and cheapest transportation method for smaller quantity of produce. It is convenient for a village or other areas where human being is the only means of transport. Bags of oilseeds are carried on the head or back of human being. Coolies are used for such purpose which is common in urban areas.

Pack animals: Suitable for carrying oilseeds in smaller quantities by animals like camel or donkeys in a village, hilly or desert tracks where other transport like bullock/ camel carts are not suitable.

Road transport

Thelas: It is two or four wheeler vehicle driven by human being and commonly used for carrying oilseeds from door to door in nearby areas.

Rickshaws: In some semi-urban and urban areas, cycle rickshaws and rickshaws are very convenient and popular mode of oilseed transport to short distant places.

Tongas: These are horse driven vehicle suitable for carrying oilseeds in narrow roads.

Bullock / camel cart: Bullock or camel carts are the primary means of transport in most rural areas of India. Capacity varies between 187kgs. to 933kgs. It is convenient due to following reasons:

— Cheap and easily available conveyance for the farmers having small quantity of produce to be transported to short distant areas.

— Operational cost is low.

— Easily manufactured by village artisans from materials (wood) available at village and the repairing facilities are also readily available there.

— No special type of road is required, can be operated on Kaccha road, muddy or sandy path also.

— Multi-purpose transport system creates employment to village artisans.

Tractor trolley: Transportation by tractor trolley is convenient due to following reasons.

— To carry larger quantity of produce than bullock carts in less duration of time.

— Suitable in surplus producing areas than the trucks for carrying produce to the primary assembling markets where there is absence of proper pucca road connecting to the villages.

Yugad: It is a kind of four wheeler which is widely used in the rural areas of Rajasthan and Uttar Pradesh states for carrying mustard-rapeseed.

Truck: For larger or bulk quantity, the truck is the most convenient mode of transport throughout the country and in some cases better than railway wagons since the railway wagon transportation poses some difficulties like timely non-availability of wagons, safety of goods and problems of loading-unloading of produce directly at godowns. Capacity of truck varies between 8-12 tonnes. It is convenient due to following reasons.

— Easy Availability
— Time saving
— Quick movement of stocks
— Door to door delivery
— Comparatively cheaper for short / medium distances
— Suitable for smaller quantities at a given time.
— Flexibility in operation and reliability in handling of produce.
— Lower transit losses due to least handling of loading and unloading.

Railway transport

Railway system in India is Asia's largest and world's fourth largest in terms of route kilometers. The railway system is developed in India in a manner so that it can transport bulk commodities for long distance at a cheaper and economical rate. But the transportation of mustard-rapeseed through railways is not common in India.

The transportation of mustard-rapeseed by railway wagons may become convenient for following reasons:

— Suitable for carrying larger quantity of produce over long distances through out India.
— Comparatively cheaper and safer mode of transport available through a wide network of railways.
— Facilities of subsidised tariff for agricultural commodities is available depending upon the situation.

During transportation, following care should be taken:

— The packages of oilseeds should be handled and transported in such a way so that they remain well protected from sun, rain or other sources of excessive heat, objectionable odour and from any type of cross infestation.

— During transportation, there should be proper arrangement of sufficient aeration and insulation to reduce the heat.

— Stacking height should be kept up to 6 to 10 tiers.

— While handling and lifting of bags during transportation, too much use of hooks by labourers should be avoided, as it may cause spillage losses.

Storage

Mustard–rapeseed is living organism and therefore, develop carbon-dioxide and water at the expense of sugar and starch. Such activities are enhanced by moisture, temperature and relative humidity. Before storage, it is essential to clean oilseeds and remove plant foliage and stems etc which initiate the heating, development of carbondioxide and quality deterioration in seed mass. Hence, the mustard-rapeseed should be stored under low moisture content (8 per cent) and temperature (25°C). In storage, the mustard - rapeseed is protected from the time of plenty during production period to the future consumption during the scarcity period.

Safe storage

The following requirements should be fulfilled for safe storage of oil seeds like mustard - rapeseed:

i) *Selection of godown:* The bags of mustard seeds should be stored in covered premises, which are well protected from moisture, excessive heat, insects and rodents. The godown should be made on a well built platform of a height of not less than 1ft. from ground level to prevent soil moisture and dampness. The roof of the godown should have sufficient height from the mustard seed stacks for keeping minimum possible temperature in godown. Sufficient space should be provided between stacks for proper air circulation in storage.

ii) *Cleaning of godowns:* For safe storage, godown should be properly cleaned so that there should be no left over grains which may harbour infestation and contaminate the new stock. The walls of godown should be whitewashed alongwith painting of walls with coal tar up to the height of 1 ½ meter. Before storage, the godown should be sprayed with Malathion or DDVP. Besides, care is to be taken also for filling up of cracks and crevices of godowns, which also harbour crawling infestation.

iii) *Cleaning and drying of oilseeds:* Before storage, the oilseeds should be properly cleaned and dried so that it should be free from dust, dirt particles, moisture which results in quality deterioration of oilseeds.

iv) *Separate storage of old and new stock:* To check crawling infestation and to maintain hygienic condition of godown, the old and new stocks are to be stored separately.

V) *Cleaning of bags:* Always new and dry gunny bags should be used for oilseeds storage. But the old gunny bags can be used only after cleaning and drying in sun and also fumigating by Alluminium Phosphide.

vi) *Cleaning of vehicles*: The vehicles used for transporting oilseeds should be cleaned with phenyl.

vii) *Use of dunnage:* Bags of oilseeds should not be stored directly on the floor of the godown. These should be kept by arranging wooden crates or bamboo mats along with a cover of polythene sheet. Otherwise, the bottom layer of oilseed bags of the stack may be damaged by absorbing moistures from the floor.

viii) *Aeration of godown:* To provide regular aeration to the stock in the godown is very essential to maintain the quality of the stock. Hence, proper aeration should be given during clear weather but care is to be taken to avoid aeration during rainy days.

ix) *Regular inspection of stocks:* To maintain health and hygiene of the stocks, regular inspections every fortnight (15 days interval) is essential.

Storage structures

— *Earthen pots:* These are cylindrical shaped structures made up of unburnt clay mixture with straw or cowdung mixture, mud and bricks. Generally used by farmers at rural areas.

— *Bamboo baskets:* Baskets made of bamboo are used by some producers in villages to store mustard - rapeseed in smaller quantity.

— *Gunny bags*: Bags made of jute are widely used throughout India by the producers, traders, processors, packers, etc., for storing of mustard seeds in larger quantities.

— *Circular steel bins (anaj kothi):* It is convenient for storing smaller quantities of mustard - rapeseed up to 3 tonnes. Assembling 4-6 pieces of corrugated or plain M.S. Sheets makes it. Useful for preserving mustard seeds at domestic level.

— *Godown:* It is a pucca constructed storage structure where traders generally store the bags of mustard-rapeseed by building the stacks in it. It is made of cemented brickworks wall and floor along with corrugated roof.

— *Warehouse:* This scientific storage structures are created and widely used by organisations like Central/State

Warehousing Corporations (CWCs, SWCs), National Agricultural Co-operative Marketing Federation (NAFED) and other co-operative marketing organisations on a large scale in different parts of India. These storage structures are always built as per prescribed guidelines for scientific storage of agricultural commodities.

Storage facilities

i) *Producers' storage:* The farmers generally use two types of storage structures, i.e., indoor and outdoor storage structures. In case of quantities up to 5 tonnes, indoor facilities can be used but for larger quantities, outdoor structures are preferred. As regards indoor structures, farmers generally prefer metal bins whereas in case of outdoor structures they prefer steel, aluminum or concrete godown /sheds.

ii) *Rural godowns:* The godowns / warehouses generally used by the farmers or village traders are not always scientifically built and maintained, which sometimes results in storage losses. However, for building up of scientific godown at the village level, the benefit of Gramin Bhandaran Yojana scheme of Govt. of India should be availed.

However, the following points should be kept in view during construction of buildings of rural godowns:

— Topography of the area of construction
— Ground condition, elevation etc.
— The foundation, floor and walls should be of sound designed concrete based structures.
— Roof should be constructed with galvanised steel having wide span and no supporting pillar.
— There should be provision of sufficient ventilation and lighting facilities.

iii) *Mandi godowns:* The Agricultural Produce Market Committee(APMC) complexes are constructed with storage facility so that the agricultural produce brought to the market should be stored safely by market committees. During storage, the goods are weighed in presence of seller /producer and receipt is issued indicating the nature and weight of goods to be stored and it is issued by the licensed general commission agents, brokers depending upon the case. In most of secondary and terminal regulated markets, Central and State Warehousing Corporations also provide scientific storage facilities at prescribed storage charges and issue warehousing receipt against pledge of produce, which is a negotiable document for obtaining finance from the scheduled bank.

According to a survey conducted by the Directorate of Marketing & Inspection (DMI), out of 89 markets, 66 markets (74per cent) were provided with storage facility and rest 23 (26per cent) were without storage godowns.

iv) *CWC & SWC warehouses :*

— *Central Warehousing Corporation (CWC):* It was established in 1957. It is the biggest public sector warehouse operator in the country. The CWC has godowns in important markets of all states. It is the largest public sector warehouse operator with their godowns established in almost all the states of the country. At present, CWC has already established and operating 464 warehouses being managed by 15 regional offices. These are scientifically constructed warehouses which facilitate the farmers to store their produce safely and to device the benefit of pledge finance for safeguarding against distress sale during the period of glut situation in the markets. The warehouses of CWC provide storage and ancillary services for more than 250 groups of commodities and products. It provides specialised arrangements and high degree of professional cares and skill for many of these products. These storage facilities are utilised for all commodities including mustard-rapeseed. All India storage capacity of C.W.C (in lakh tonnes)

— *State Warehousing Corporations (SWCs):* The Central Warehousing Corporation (CWC) has 17 associates in State Warehousing Corporations (SWC) which were established in different states and located at distant places. The total share capital of SWC's is contributed equally by CWC and the concerned state governments i.e. it is under dual control. The SWCs also provide storage facilities for oilseeds like mustard -rapeseed.

v) *Co-operative storage:* In Co-operative sector, National Co-operative Development Corporation (NCDC) has been making systematic efforts to assist to establish scientific storage facilities in co-operative sector. Till the end of March, 2001, NCDC increased it's financial assistance to build storage capacity upto 137.67 lakhs tones. The corporation has been implementing it's storage programmes under the following schemes:

— Centrally Sponsored Schemes.

— Corporation Sponsored Schemes.

— EEC assisted Rural Growth Centre Project.

Pledge finance system

The Indian farming community mostly consist of small and marginal farmers. They do not have the economic strength to retain the surplus produce till favourable market price and often compelled to sell their produce immediately after harvest when the prices are low. The solution to this problem lies in providing safe and scientific storage of their produce and availing easy marketing credit against the stored produce. Hence, the system of pledge finance have emerged as an unique avenue of finance to farmer.

REFERENCES

Chand, Roja, *Agri Clinic: A boon to Farming, Agro India,* Vol.III, Issue 4, 2003, pp-12-14.

Gantzer, Hugh & Collean, "Mustard: The Versatile Spice," *Indian Spices,* Vol.35, No.2, 1998, pp.8-9.

Gupta, Shakuntala and Balwinder Singh, "Price Spread in Marketing of Groundnut and Rapeseed-Mustard in Punjab," *Indian Journal of Agril Mktg.*:12 (1 & 2), 1998, pp.128-136.

Kumar, Arbid and J.S.Chauhan, "Challenges in Rapeseed – Mustard Production," *SAARC Oils and Fats Today,* Vol.V, Issue 7, 2003, pp.32-33.

Motey, Rashmi and Smita Lele, "Plastic Films for Processed Foods – Special Requirements," *Packaging India,* Vol.35, Issues No.-5, 2003, pp-19-31.

Mohd, Vandan, "Marketing Extension and Commercial Farming," *Yojana,* July 1999, 1999, pp-37-39

Ram, Kewal, *Saarc Oils & Fats Today,* Vol – V, Issues – 11, 2003, pp-28-30.

8

Value-added Utilisation of Oilseeds

The oilseed industry is a vast, worldwide enterprise. Nearly every country in the world produces oilseeds of one kind or another and most of them have at least a rudimentary oilseed processing industry that produces two basic foods: edible oil and protein meal. Edible oils are consumed directly in the human diet in products like cooking oil, salad oil, margarine and shortening. They also are an ingredient in many food products and industrial products.

The protein meal that remains once the oil has been extracted from the oilseed is used primarily as a feed ingredient for poultry, hogs, dairy cattle, beef cattle, lambs, fish and other animals, including pets. In essence, protein meals are used to produce animal proteins in the forms of meat, milk and eggs. Other uses of oilseeds and their derivatives include soy foods as well as pharmaceuticals and industrial products. Vegetable oil also can be feedstock for the production of methyl esters, commonly called biodiesel, that can be used as a fuel in diesel engines.

Although the primary uses of oilseeds are related to food production, oilseed production has grown much more rapidly than the growth in world population over the past 20 years. This indicates a rising standard of living worldwide. As the standard of living rises, so does demand for a better diet featuring more vegetable oil and meat, milk and eggs. Had world *oilseed* production increased directly with world population increase, production of oilseeds in 2005 would have been 258.3 million MT rather than 371.9 million MT. Actual production exceeded this amount by 44%.

Table 1. World Oilseed Production and Population Selected Years, 1990 – 2005

Year	*Production of 10 oilseeds*	*Soybeans %*	*World population in of all oilseeds millions*
2005	371,940	57	6,455
2004	329,930	56	6,379
2003	328,250	60	6,302
2002	321,930	57	6,225
2001	311,800	57	6,148
2000	302,280	53	6,070
1995	260,560	53	5,674
1990	213,640	50	5,338
Percent increase 1990 to 2005	74%	97%	21%

Had world *soybean* production increased directly with world population increase, production of soybeans in 2005 would have been 130 million MT rather than 212 million MT. Actual production exceeded this amount by 63%. These data dramatically demonstrate the ever-increasing role of oilseeds in world food production, especially soybeans, due to the unique qualities of soybean meal as a protein source in animal rations.

ROLE OF THE UNITED STATES

While the oilseed industry continues to expand rapidly worldwide, the role of the United States has diminished over the past 15 years in both oilseed production and processing. While oilseed crush in the rest of the world increased by 50 million MT (25%) between 2000 and 2005, oilseed crush in the United States was virtually flat during that period. Soybeans are by far the dominant oilseed grown and processed in the United States, accounting for 88% of oilseed production during the past five years.

Although the U.S. continues to lead the world in soybean production and processing, those leads are declining. The United States' share of world soybean production has declined from 50% in 1995 to 40% in 2005. The United States' share of world soybean crush has declined from 35% to 26% during the same period. The primary shift in world soybean production has been from the United States to South America.

Ten years ago China processed 7.8 million MT of soybeans and used 6.4 million MT for food, feed and seed. Of the 14.2 million MT of soybeans consumed by China in 1995, all were produced domestically and China was a net exporter of 0.3 million MT of soybeans. China produced 76% of its edible oils and fats and 103% of its protein meals in 1995, i.e. China was a net exporter of protein meals, primarily soybean meal (900,000 MT).

Five years ago China processed 16.9 million MT of soybeans and used 7.7 million MT for food, feed and seed. Of the 24.6 million MT of soybeans consumed by China in 2000, 62% were produced domestically with the remaining 38% being imported. China produced 85% of its edible oils and fats and 99% of its protein meals in 2000. In 2005 China will process 29.7 million MT of soybeans and use 9.4 million MT for food, feed and seed. Of the 39.1 million MT of soybeans that will be consumed by China in 2005, 45% will be produced domestically with the remaining 55% being imported.

China will produce 71% of its edible oils and fats and 100% of its protein meals in 2005. (China will import and export about 1.3 million MT of protein meals in 2005, with the primary imported meal being fish meal at 1.1 million

MT and the primary exported meal being soybean meal at 880,000 MT.) In the case of Argentina, the increase in soybean processing is related to a greatly expanded production of soybeans, the products from which have become a significant earner of foreign exchange for the country.

Export duty structure favours processing beans within the country and exporting the products, i.e. the value is added within Argentina. Ten years ago Argentina produced 12.5 million MT of soybeans, processed 8.7 million MT or 70% of the crop and used 0.5 million MT for food, feed and seed. Argentina was an exporter of 2.5 million MT of soybeans, or 20% of the crop. Argentina exported 97% of the soybean oil that it produced in 1995 and 93% of the soybean meal.

Therefore, even ten years ago Argentina had established itself as a large soybean producer that processed most of its crop and then in turn exported most of the products. However, the size of Argentina's role in supplying the world with soybean oil and soybean meal was nowhere near the size that it has become today. Five years ago Argentina produced 21.2 million MT of soybeans, processed 16.5 million MT and used 0.8 million MT for food, feed and seed. Argentina was an exporter of 4.1 million MT of soybeans, or 19% of the crop.

Argentina exported 93% of the soybean oil that it produced in 2000 and 98% of the soybean meal. Even with crush increasing by 77% during the five years exports of both oil and meal kept pace, i.e. Argentina continued to establish itself as a primary supplier of soybean oil and soybean meal to the world. In 2005 Argentina produced 38.7 million MT of soybeans. It will process 26.8 million MT (69% of the crop) and use 1.5 million MT for food, feed and seed. Exports will be 9.1 million MT or 24% of the crop.

Argentina will export 93% of the soybean oil that it will produce in 2005 and 97% of the soybean meal—virtually the same percentages as five years earlier even as crush increased by 68%. Indeed, Argentina will supply 48% of the soybean oil that will trade internationally in 2005, far outdistancing Brazil (28%) and USA (7%). Argentina will supply 44% of the soybean meal that will trade internationally in 2005;

Huge plants of 8,000 to 10,000 T/day capacity have been built along the Rio de la Plata to take advantage of the abundant supply of raw material, low operating costs and direct loading of oil and meal from the plant to a ship at the plant's loading berth. There is no question that Argentina has established itself as the primary supplier of soybean meal and soybean oil to the world.

ROLE OF MINNESOTA

The only oilseed of which Minnesota produces a significant quantity is soybeans. Soybeans are by far the #1 oilseed crop grown in Minnesota, representing over 99% of the total oilseed crop. Indeed, other oilseeds are falling more and more out of favour with Minnesota growers. The traditional growing area for sunflower, canola and flax, the Northwest Agricultural Statistics District, has seen a switch to soybeans in recent years as new varieties have been developed for extreme northern growing areas.

During the 8-year period, soybean acreage increased by 148% in the Northwest District while it was virtually unchanged in the rest of the state. The Northwest District accounted for 17% of soybean acres in Minnesota in 2005 compared with 7% in 1997.

The five soybean processing plants in Minnesota have total capacity of 14,875 MT/day, enabling Minnesota to process 60% of its soybean crop, based on operating at capacity for 310 days/year. Both Iowa and Illinois, the top two soybean producing and processing states, have crushing capacity equal to about 75% of the crop. Number

4 producing state Indiana can crush about 60% of its crop; number 5 Ohio about 50% of its crop. Total USA soybean crushing capacity equals about 65% of the crop.

The purpose of the 2000 study was to examine a wide range of ways to add value to oilseeds. It was not intended to be an in-depth analysis of any of the possibilities that were examined. The intent was to identify those possibilities that looked positive enough to warrant further study.

The following opportunities were examined in the 2000 study, along with the recommendations made at that time:

— Commodity soybean processing: not recommended
— Identity-preserved (IP) soybean processing: investigate further if a grower's group can be formed that is willing to commit the required time and effort
— Soybean mini-mill: investigate further
— Multiseed processing, including corn germ: investigate further
— NuSun™ sunflower oil: allied with multiseed plant
— Press vegetable oil: allied with mini-mill or multiseed plant
— Derivatives of soybean oil: investigate further
— Soy concentrates and isolates: not recommended
— Soy flour: not recommended
— Isoflavones: investigate further, perhaps with Schouten USA
— Oligosaccharides: not recommended
— Soy-based foods: not recommended, except perhaps soy milk
— Salad dressings and sauces: not recommended
— Biodiesel: not recommended, unless allied with an IP plant or mini-mill

— Industrial uses of soybeans: not recommended, unless allied with an IP plant or mini-mill

Each of these opportunities will now be revisited, including how low linolenic acid soybeans can address the growing health concern over trans fatty acids, and examination of the role and availability of rendered fats and greases and waste oils and fats in the biodiesel industry.

Commodity Soybean Processing

Commodity soybean processing refers to the basic industry that extracts the oil from the bean using solvent-extraction technology. Both products, the oil and the meal, are commodities, that is, there is no significant difference between one processor's products and those of another processor. At the time of the 2000 study, both CHS and MnSP had already announced plans to build 2725 MT per day plants in Fairmont and Brewster, respectively. Both were built and both started up in late 2003.

As pointed out in the 2000 study, this was too much additional crush capacity for Minnesota to absorb at once even though it brought total crush capacity in Minnesota into line with other large soybean producing states, as measured by the percent of the crop that can be crushed. There are two main reasons why this is so. One is the geographic position of Minnesota within the soybean crushing industry in the United States. Being located at the northern edge of country means that Minnesota processors must go past other processors when moving products out of the state.

The only non-Minnesota market that Minnesota crushers are closer to than other crushers is north into Canada, and demand for soybean meal and oil is very limited in Canada due to the country's small population, most of which is located far to the east of Minnesota anyway. There is not enough consumption of oil and meal in Minnesota to allow all the products to be consumed within the state. This is especially a problem with meal, which tends to be a more local market than oil.

The 2000 study pointed out that soybean meal consumption in Minnesota was nearly in balance with the existing industry's meal production capacity. At that time the number of poultry and livestock units in the state indicated soybean meal consumption of about 1.7 million MT/year, with soybean meal production capacity of about 2.2 million MT/year.

During the past five years animal units in Minnesota have changed little. Increases in hog and turkey production have been partially offset by reductions in dairy cattle. Overall soybean meal consumption has increased by only 55,000 MT/year while soybean meal production capacity has increased by 1,267,000 MT. This means that Minnesota now has the capacity to produce nearly 3.5 million MT/year of soybean meal while consumption by livestock and poultry within the state is only 1.8 million MT/year. The surplus 1.7 million MT/year must find homes outside of the state.

Surplus meal production capacity in the state has increased from 464,000 MT (27% of meal production capacity) five years ago to 1,676,000 MT (94% of meal production capacity) today. The only practical way to consume a significant amount of Minnesota's soybean meal production capacity surplus within the state is more animal units on feed. And even if there were economic justification for large expansion of animal units in Minnesota, it would be a battle to receive the required permits to do so.

The second reason why it is difficult for Minnesota soybean processors to compete for markets outside the state, in addition location, is the fact that Minnesota soybeans are lower in protein content than those grown in the major soybean producing states from Nebraska to Ohio. Minnesota processors generally guarantee a protein content of 46.5% on hipro (dehulled) soybean meal while the industry standard is 47.5% to 48%. If 47.5% protein meal can be bought for $225/T, 46.5% protein meal is worth only $220/T. This translates into a difference of about 11 cents/bushel in the margin for the processor..a very significant difference in an industry which measures its margin in cents/bushel.

On the oil side, based on average USA annual soybean oil consumption of 60 pounds per capita and a population of 5 million people, Minnesota consumes 136,000 MT of soybean oil per year in the forms of cooking oil, salad oil, margarine, shortening, and food ingredients. The soybean oil production capacity within the state is about 850,000 MT/year. While the oil side is even more disproportionate than the meal side, the two cannot be compared directly.

The market for soybean meal is much more localised than that for soybean oil. Oil is often shipped long distances to the customer and Minnesota soybean oil is no different from other soybean oil, unlike the situation with the lower protein content of Minnesota soybean meal. In addition, soybean oil must be refined before it becomes an edible product. The ADM plant at Mankato is capable of refining all its crude oil production.

The CHS refinery at the Mankato crush plant can refine all the crude oil production from both the Mankato plant and the new Fairmont plant. CHS uses this large refining capacity to serve its Ventura Foods line of vegetable oil products. The MnSP plant at Brewster does not have a refinery but it started up a biodiesel plant this summer that can take up to two-thirds of its oil production. The AGP plant at Dawson has no refinery or biodiesel plant, making it the only processor in Minnesota that does not add value to its oil but must ship it to other refiners as crude oil.

In addition to the generally unfavourable situation in which to add soybean crushing capacity in Minnesota, the timing of the start-ups at Fairmont and Brewster was not good, as the soybean crop in 2003 was down considerably from the record crop of the previous year due to drought conditions during the growing season, from 8.4 million MT (309 million bushels) in 2002 to 6.5 million MT (238 million bushels) in 2003. Things did not improve in 2004, with a crop of only 6.4 million MT (236 million bushels). However, the 2005 soybean crop was forecast to be 7.8 million MT (286 million bushels) in the October 12 crop report.

One processor stated that crushing margins have been "terrible" in Minnesota the past two years due to the small crops and the need to ship soybean meal past existing processors farther south in order to find homes for all the meal. Gross crushing margins often have been in the 30-35¢ per bushel range; full operating costs exceed this by 20¢ per bushel or more.

Minnesota has excess soybean processing capacity for the foreseeable future. The soybean processing industry is mature in the United States. Countries that formerly were big importers of meal and/or oil are now producing it themselves. The historical trend is that soybean crush moves to where the products are consumed. This happened in the United States 35 years ago when considerable crushing capacity was built in the Southeast, to serve the large poultry producers there. It has happened in China over the past five years.

It is taking place in the Middle East today. Europe crushes 33 million MT/year of oilseeds while producing only 20 million MT. The US has become the residual supplier of soybean oil and meal to the world, meaning that any expansion of the industry within USA must be for the purpose of supplying domestic markets for meal and oil, which also are mature, more or less growing only at the rate of population, which is only 1% per year.

IP Soybean Processing

The purpose of identity-preserved (IP) processing is to

segregate one or both of the products due to specific traits of the product which make it something other than a commodity. In order to do this a processor must be capable of IPing the beans as well as the desired product(s). This is not always easy to do since most soybean processing plants are configured only for commodity processing, not IP processing. Special trait soybeans have been under development for at least the past decade.

Most of these are genetically modified (GM) to give the soybean the desired trait. The first such soybean to become commercially available was the herbicide-resistant Roundup Ready™soybean developed by Monsanto in the mid 1990s. It rapidly gained favour with USA (and Argentine) farmers to the extend that an estimated 87% of the USA soybean crop in 2005 had the Roundup Ready™ characteristic, while 99% of the Argentine crop did. Roundup Ready™ soybeans accounted for an estimated 83% of the Minnesota soybean crop in 2005.

Herbicide resistance is an example of an agronomic trait as opposed to a nutritional trait. Many people believe that the products (oil and meal) from GM soybeans are no different from those from non-GM soybeans. Others believe that products from GM soybeans, especially the meal (since the GM trait is detectable only in the meal) could be harmful to animal and/or human health.

The concern over health issues related to GM foods is greatest in Europe and Japan. At least one USA commodity processor, Consolidated Grain and Barge in Mount Vernon, IN, has been offering non-GM soybean meal to Japanese customers at a premium to non-GM meal for several years. This is an example of IP processing. Although to date special trait soybeans have been predominantly agronomic ones, there also are a number of nutritional trait soybeans under development.

The trait can pertain to either the oil or the meal and can be for either animal or human nutritional benefit. On the oil side special traits include low saturated fat, greater resistance to oxidation, greater heat resistance, longer shelf life and biodegradable plastic production. On the meal side they include more complete digestion in animal feeding, cancer and osteoporosis prevention, prevention of meat-borne diseases and reduction in off-flavour in soy-based foods.

An early example of IP processing of a soybean with a special nutritional trait was the LoSat™(low saturated fat) soybean developed my Pioneer Hi-Bred and processed by Zeeland Farm Services (ZFS) of Zeeland, Michigan. The initial arrangement was that Pioneer Hi-Bred retained ownership of the beans and products and paid ZFS a fee to process the beans in what is known as a toll crush arrangement. However, this proved not be effective, so now LoSat™beans are grown and processed like any other, with the grower buying the seed at a premium and selling the crop to ZFS at a premium, who processes it and markets the products, receiving a premium for the oil.

IP processing should continue to expand with the development of more and more special trait soybeans. Some existing processors already are doing this and others may follow suit. This also would be a strategy to pursue by any new processors, particularly mini-mills, which by their nature usually are better suited to IP processing than are the large commodity processing plants.

Linolenic acid is one of the many fatty acids contained in soybean oil. Typically soybean oil contains about 8% linolenic acid. Monsanto has developed, through conventional breeding, a soybean that produces oil with only about 3% linolenic acid content. This gives the oil greater stability and eliminates or reduces the need to hydrogenate the oil to accomplish greater stability. Like saturated fats, hydrogenated oils contain trans fatty acids (trans fats) that can raise the level of LDL (bad) cholesterol and lower the level of HDL (good cholesterol), a major health concern to many Americans.

As of January 1, 2006, listing of trans fat content will be required in nutritional labelling in the United States. The development of low linolenic acid soybean oil will be beneficial in addressing the trans fat labelling issue. There is little doubt that trans fats have become and will continue to be a health concern. In August a judge approved an $8.5 million educational campaign by McDonald's to settle lawsuits for failing to reduce its use of partially hydrogenated cooking oil. The campaign includes a payment of $7 million to the American Heart Association to educate people about trans fats and $1.5 million to publicize that the company had not followed through on a pledge made in 2002 to switch to more healthful cooking oil.

Recently a Cleveland-area fast food chain opted to use trans fat-free canola oil in all its restaurants, no doubt as a way to promote its food as being more healthful. Others certainly will follow suit as the general public becomes more aware of the health risk posed by trans fats. VISTIVE™ soybean acreage was small in 2005 but Monstanto expects that acreage will increase to nearly 500,000 acres in 2006.

In press releases dated August 10, September 6, September 26 and October 6, 2005, Monsanto announced that CHS will contract for up to 40,000 acres of VISTIVE™ soybeans in southern Minnesota and northern Iowa in 2006, ZFS will contract for up to 25,000 acres in Michigan, AGP will contract for up to 150,000 acres in Iowa and Cargill will contract for up to 150,000 acres in Iowa, bringing total contracted acres for 2006 to 356,000, as of this writing.. In all cases, the processor will pay a premium for contracted VISTIVE™soybeans, and then it will crush and sell the processed soybean oil to food company customers.

Presumably the processor will be able to receive a premium for low linolenic acid soybean oil; ZFS confirms that this is indeed the case. Monsanto stated that similar agreements will be made by Monsanto with other processors in order to reach the nearly 500,000 acres that Monsanto anticipates being planted in 2006.

Pioneer Hi-Bred has been a leader in development of nutritional special trait soybeans. In addition to their LoSat™ soybeans mentioned previously, they also are ready to commercialise low linolenic acid soybeans. They anticipate 200,000 acres in 2006. Since Pioneer Hi-Bred is owned by DuPont, and DuPont is in partnership with Bunge in what they call the Bunge DuPont Biotech Alliance, their program will be different from Monsanto's. Bunge will contract for all the acres and process the soybeans, with the Alliance sharing production costs and profits.

Since Bunge has no processing plants near Minnesota, their program will not be a factor here, at least not until the acreage of low linolenic acid soybeans is much larger than it is today. Pioneer Hi-Bred is working on many other varieties of special nutritional trait soybeans. Many of these are beans in which the special trait resides in the meal, either for improved animal nutrition or nutraceuticals for humans. No doubt these soybeans will be kept under the control of the Bunge-DuPont alliance, which will incorporate the meal into their Solae™ line of soy foods and nutraceutical products.

Soybean Mini-mill

This continues to be an intriguing concept. Although the United States commodity soybean processing industry is mature, there still are opportunities for small processors to fill niches for specialty processing that are not of interest to the large processors.

There are a number of operating mini-mills scattered across the country. An example is Thumb Oilseed

Producers (TOP) of Ubley, Michigan. TOP is a "new generation" cooperative with a small (125 T/day) expeller plant. The members grow only organic and non-GM soybeans. From them TPO produces Soy Beginnings® brand of non-GM soy oil. The equipment in the plant (extruder/expeller and refinery) are such the oil can be physically refined, that is, it does not require the chemical treatment that is needed to refine soybean oil produced by the conventional solvent extraction process.

TOP also is a partner with Spectrum Foods of Springfield, Illinois, and InstaPro International of Des Moines, Iowa, in a venture called NexSoy, which markets NexSoy™ brand of organic soybean oil. InstaPro International makes the equipment that processes the soybeans and refines the oil; TOP is the processor/refiner and Spectrum Foods is the marketer.

Another expeller plant that has enjoyed considerable success for many years is West Central Soy (WSC) at Ralston, Iowa. WSC can hardly be called a mini-mill today since it has increased its capacity to 1000 T/day over the years. WCS originated as a dairy cooperative producing soybean meal with bypass protein that translates into more milk production than with normal soybean meal in the ration. WCS's success with bypass protein soybean meal is such that it ships its meal all over the United States at a premium to normal soybean meal of about $15/T.

In fact, WCS's meal is so valuable that it ships none of its meal to its own feed mill next door to the soybean processing plant, but rather buys solvent extracted meal from other processors. WCS also has been a biodiesel producer for several years. A couple years ago it installed a new biodiesel plant with capacity of 12 million gallons of biodiesel per year, capable of consuming all its soybean oil production. Until recently this was the largest biodiesel plant in the United States.

Key to the success of a mini-mill is making a specialty product that is not produced by the large commodity processors. In the case of some mini-mills this is the oil; in the case of others it is the meal. But at least one of them must be special and command a premium to commodity oil or meal, or the plant will not be able to compete, since the cost of operating a mini-mill is greater than that to operate a large solvent extraction plant on a per unit basis. In addition, the oil yield from an expeller plant is lower, and oil is the higher-valued of the two products.

Mini-mills do have some advantages over solvent extraction plants, though. Since their capacity is smaller, they do not require as large a capital investment, environmental permitting is easier, operation is simpler and safer and expansion is simpler. Also there is new expeller technology being developed using supercritical CO2 that has the potential to extract more oil than conventional expellers are able to as well as produce meal/ flakes with a higher protein dispensability index (PDI) that is needed in producing soy foods.

Mini-mills also are well-suited to IP process specialty crops that are not produced in great enough volume to attract the large commodity processors. There is a cooperative, Central Minnesota Soybean Processors (CMSP), centered in Otter Tail County, which is considering building a mini-mill in the Ottertail area. There should be a ready supply of soybeans for the plant, given the increase in soybean production in this part of Minnesota during the past few years.

The question remains whether or not the plant will be able to produce a specialty product and have an adequate market for such a product. There should be assurance that this will be the case before investing in this plant. There is a large market in central and west-central Minnesota for soybean meal due to the large turkey and dairy cattle populations in the region. However, a mini-mill would have a difficult time competing with the commodity processors in Enderlin and Dawson for these markets unless they can develop a premium market for their meal, as WSC has done, primarily serving the dairy industry. On the oil side, the plant could pursue a strategy either of being a supplier of "natural" oil, like TOP, or biodiesel production.

Multiseed Processing

At the time that the 2000 study was conducted, this appeared to be worth further study. The idea was to locate the plant in west-central Minnesota to be close to the sunflower producing areas of North and South Dakota, the minor oilseed producing area of Minnesota and the Minnesota Corn Processors (MCP) plant at Marshall, which does not have a germ processing plant.

However, things have changed considerably since then. The biggest change is the fact that ADM now owns MCP. ADM has its own corn germ processing plants and would not be a supplier to a multiseed plant that would be in direct competition with its multiseed plant at Enderlin. Corn germ from MCP would have been an important feedstock for the multiseed plant.

In addition, it was assumed that canola production would increase in Minnesota, following the pattern that occurred in North Dakota, where canola acreage increased from 200,000 acres in 1996 to 1.1 million acres in 2000, But this did not happen. Minnesota canola acres declined from 140,000 in 2000 to 30,000 in 2005. Certainly there is no room for a large new multiseed plant in Minnesota at this time.

NuSun™

NuSun™ is the trade name for mid oleic acid sunflower oil. NuSun™, like low linolenic acid soybean oil discussed previously, is a non-trans fat oil. It has caught on well since its introduction five years ago or so to the point where the National Sunflower Association estimates that 85% of the oil sunflower crop in 2005 is either high oleic or mid oleic, with most of this being mid oleic. The crushers pay a premium for NuSun™ of about 5% over regular oil sunflower.

Production of NuSun™ sunflower oil was seen as a strategy for a multiseed plant in west-central Minnesota to pursue. Since that no longer looks like a viable opportunity, production of NuSun™ sunflower oil in Minnesota has become a moot point at this time.

Press Vegetable Oil

Press (or expeller) vegetable oil is marketed by "natural" food companies at a significant premium to solvent extracted oil. The two major players in this market have been Spectrum Organic Products of Petaluma, California, and Hain Celestial of Melville, New York. Neither company has its own vegetable oil production plant but have their oil contract manufactured for them, mainly by expeller plants in California.

Both Spectrum and Hain indicated during discussions with them two years ago that they would entertain the idea of having some of their oil produced somewhere nearer the markets east of the Mississippi River in order to save freight costs..both freight to get the oilseeds to California to be processed and to ship the product back to the eastern half of the United States for distribution.

The 2000 study favoured this strategy being pursued by a multiseed plant in west-central Minnesota since the greatest demand is for press canola oil. As with NuSun™ sunflower oil, this is now a moot point. However, it would be a strategy for a mini-mill to pursue..either for soybean oil but also for canola oil, if the plant were configured to be a multiseed processor.

As noted previously, in order to prosper a mini-mill must produce a specialty product. This could be bypass soybean meal for the dairy industry or it could be press oil for the natural food companies. If the latter, a long term supply agreement should be reached before proceeding with construction of the plant. If CMSP proceeds with its plan to build a mini-mill in west-central Minnesota, an alliance with Hain should be investigated. It will be important to do this prior to committing to the type of equipment to put in the plant.

Derivatives of Soybean Oil

This opportunity was considered to be worthy of further investigation because of the high value of the products which are used in pharmaceuticals which can be derived from soy lecithin, the ready supply of lecithin from the soybean oil refineries at Mankato and the fact that no one was already making the products in Minnesota. In fact, all high concentration (30% purity and higher) phosphatidylcholine and phosphatidylserine used in the United States is imported from Europe.

Lecithin fractionation was studied further and was eventually pursued by SoyMor, a cooperative located in south central Minnesota. Sufficient capital was raised and the plant was built in Glenville, near Albert Lea, adjacent to an already existing dry corn milling ethanol plant, Exol. The SoyMor lecithin fractionation plant started operating early in 2005 and as of this writing is still fine-tuning the process before going into full scale production.

SoyMor's website states that the company will use only lecithin that has come from soybean oil that was extracted without the use of hexane. It also states that SoyMor will "deoil" the lecithin without using acetone so as to produce "toxic free" products that SoyMor believes will be command a premium in the market. One reason SoyMor pursued lecithin fractionation is because the grade of lecithin derivatives it will make are not currently produced by anyone in the United States but are imported from Europe. Thus SoyMor will have a location advantage over its European competitors and will not be competing with domestic producers of the products that it will make.

The business is too new to say whether or not it is likely to succeed. It will be very interesting to follow the company's progress. Other derivatives of soybean oil include free fatty acids and deodorizer distillate. Free fatty acids have very little value and are usually sold as a by-product by the refiner to soap companies. Deodorizer distillate can be converted to vitamin E, the market for which became saturated after ADM and Cargill each built large vitamin E plants. So the processing of deodorizer distillate is not advocated.

Soy Concentrates and Isolates

Soy concentrates (about 65% protein content) and soy isolates (about 90% protein content) are used primarily as food ingredients. They are made from soy flakes, which are the same thing as soybean meal, except for being desolventized by a method that leaves the flakes with a high protein dispensability index (PDI). This makes the protein useful as a food ingredient.

There already are many producers of soy concentrates and isolates, including all of the "Big Three," ADM, Bunge and Cargill. These companies have moved away from being purely commodity processors and have moved into the area of food ingredients, even to the extend of branding their products. Among the brands are ADM's NutriSoy® line of "100% natural soy protein" and Cargill's Prolisse® soy protein, promoted "for heart health and overall nutrition," a product of Cargill Health & Food Technologies (CH&FT).

Recently CH&FT was included in the "Companies to Watch" list of Nutraceuticals World. The Bunge-DuPont alliance product, Solae™ was mentioned previously. The soy protein business is very competitive. CHS, which produces soy flakes at its processing plant in Mankato, has examined soy isolate production in the past and has decided against it. Rather, it grinds some of its flakes into soy flour and sells the rest to other producers of soy flour, concentrates and isolates. As in the 2000 study, the soy concentrate/isolate business is not recommended.

Soy Flour

Soy flour is another food product that is made from soy flakes. Besides its high-protein nutritional benefit, soy flour also has functional characteristics which make it advantageous to incorporate it into some food products that also use wheat flour, such as bread. As with soy concentrates and isolates, it is advantageous for a soy flour producer to produce its own soy flakes. In addition to CHS, ADM, Bunge and Cargill also are soy flour producers. With limited demand for soy flour in Minnesota, it does not appear that there is a need for additional production within the state at this time. If there were, CHS would be in the best position to provide it by expanding their existing soy flour capacity.

Isoflavones

Isoflavones, technically called phytoestrogens, are compounds found in soybeans that have been shown to act like the female hormone, estrogen, in the human body. The primary use of isoflavones is to ease the physical effects of menopause. They also are thought to prevent cardiovascular and bone diseases, such as osteoporosis, and perhaps even some types of cancer. ADM is the primary producer of isoflavones in the United States, at Decatur, IL. ADM markets its isoflavones under the Novasoy™ label.

Another marketer of isoflavones in the United States is Acatris, a Dutch company within the Royal Shelton Group. Acatris has its American regional sales office in Minneapolis. When the study was done in 2000 the name of the company was Schouten. At that time Schouten's method of operation was to ship soybeans to the Netherlands from the United States or South America where isoflavones were produced from the germ of the soybean and shipped from the Netherlands throughout the world.

At the time of the last study it was recommended that Schouten (now Acatris) be approached about the possibility of producing some of its isoflavones in Minnesota, thus saving the cost of shipping soybeans all the way to the Netherlands only to have some of the product shipped back to the United States. Schouten expressed interest in at least discussing this concept.

Things have changed since then. Acatris now ships only the soybean germ to their plant in the Netherlands rather than the entire soybean, thus saving a considerable amount of freight, due to the 1:400 ratio of germ to whole soybean mentioned above. Acatris gets its germ in the United States from a facility in Iowa, the name and precise location of which they were not willing to disclose. However, Acatris did say that this relationship works well because another customer of the facility wants the degermed soybeans, probably for the production of soy milk.

When asked if Acatris might need another soybean degerming facility, in Minnesota, they said that it is not likely, since the growth in demand for isoflavones has slowed considerably since Acatris first got into the business in 1995. They attribute the decline in growth to a parallel decline in the growth of the soy foods industry in general and to conflicting studies as to the efficacy of soy isoflavones in mitigating the effects of menopause and disease prevention. Acatris has recently pioneered another health product, flax lignan, from the hulls of flax. They have flax dehulled for them at a facility in North Dakota and ship the hulls to the Netherlands to be processed.

Oligosaccharides

Soy-based oligosaccharides, a subset of inulin, have largely been replaced by inulin derived from chicory root. Food grade inulin is a white powder with a slightly sweet taste-10% of the sweetness of sugar-that is used in fatfree foods to provide flavour and texture. Since oligosaccharides derived from soybeans do not contain any quality advantage, they command a low price if marketed as a dietary fiber, of which there are many inexpensive sources.

Lacking unique qualities, a dietary fiber can compete only if it can be produced very inexpensively, i.e. the raw material must be very inexpensive and the extraction process must be simple and inexpensive. Oligosaccharides derived from soybeans do not meet these criteria. In addition, soy-based oligosaccharides also have the disadvantage of producing intestinal gas.

Soy-based Foods

Soy-based foods have been available in one form or another for a very long time. One of the first soy-based foods was texturized soy protein (TSP) which is soy flakes which have been extruded to give the product a fibrous texture. TSP has been used in many forms: small chips flavoured like bacon or cheese, hamburger extender, nuggets used in pet foods, patties that are flavoured to taste like hamburger or sausage, soy hot dogs, etc. In these uses TSP is known generally as a meat analogue.

Meat analogues have never caught on with the general public since the flavour and "mouth feel" fall short of those of meat. Today meat analogues are used primarily by vegetarians, often in the form of "veggie burgers" that are produced by many companies and sold in venues ranging from restaurants to ball parks to grocery stores and mega-discounters.

Other forms of soy-based foods are tofu, miso and tempeh. These have been a mainstay of Far Eastern diets for centuries. These foods have been introduced to the Western diet but have gained a limited following, much as the case with TSP. Soy-based dairy products and beverages, most notably soy milk, have gained greatly in popularity, especially among younger people. Not only do they appeal to vegetarians but also people with lactose intolerance.

Growth in the soy-milk market has been very impressive, from $1.5 million in retail sales in 1980, when the industry was in its infancy, to $216 million in 1998 and $745 in 2004. This represents an annual growth rate of 23% between 1998 and 2004. But growth has begun to decline, with projected sales of $1 billion by 2010. This would represent an annual growth rate of only 5% the rest of the decade.

A recently conducted study on the soy foods industry by SoyaTech of Bar Harbor, ME, entitled "Soyfoods: The U.S. Market in 2005," states that the robust growth that the soy foods industry had experienced since its inception appears to have peaked. Overall sales of soy foods in 2004 hit $4 billion, an increase of only 2.1% over the prior year. This followed an average annual growth rate of 11.8% the prior five years and was the slowest growth for the industry as a whole since the early 1980s.

In fact, four of the five top categories of soy foods had negative sales growth in 2004. Only soy milk showed a modest increase. The reasons cited for the decline in growth in the soy foods industry are "..numerous news articles and reports..recently questioning the health benefits of consuming soybased food products...

[and]..boredom by the consumers who regularly eat soy products." In other words, the market is in need of new products to recapture the consumer's interest.

Soynuts are whole soybeans that have been roasted, like other snack nuts. Five years ago the market for soynuts was very small..a total of 6-8 million pounds/year. Most were produced in Minnesota. Dahlgren and Company of Crookston, primarily a confectionary sunflower company, was the #1 producer at 3 million pounds/year. Dahlgren does not have its own brand but produces soynuts for others with brands.

Sun Valley Products of Fargo was #2 at 2 million pounds/year. American Import Company (Amport) of Minneapolis was #3 at 1 million pounds/year. Waymouth Farms of New Hope, MN, markets soynuts under their Good Sense™label. Their soy nuts were produced for them by Dahlgren or Sun Valley Products. Waymouth Farms stated that soynuts were a new and growing product for them in 2000 and growth has been impressive.

Today they offer six choices of roasted soynuts: salted, unsalted, barbecue, hot & spicy, dry roasted and yogurt-coated. But soynuts are a very small sliver of the roasted nut business. Snack peanut consumption is about 350 million pounds/year or 40 times that of soynuts. All in all, the climate in the soy foods industry has cooled off substantially while the industry awaits the "next big thing" to renew consumer interest.

Salad Dressings and Sauces

Vegetable oil is a major ingredient in salad dressings and sauces. Due to its relatively small population, not much of this type of product is produced in Minnesota. Rather, most of it is produced near large metropolitan areas like New York, Chicago and Los Angeles. It is less expensive to ship the ingredients to the place where the finished product is made than it is to ship the finished product long distances, since the packaging also must be shipped. Imagine a railroad tank car of refined soybean oil being shipped from Minnesota to a salad dressing producer in, say, Chicago, with a small portion of the finished product being shipped back to Minnesota via truck, compared with shipping most of the bottled finished product from Minnesota to Chicago.

The freight would be far greater in the latter situation. ADM and CHS have refineries at their soybean processing plants in Mankato. The ADM refinery is sized to fit the crush plant and the CHS refinery is sized to fit the total crushing capacity of their plants in Mankato and Fairmont. This is as far in the "value chain" that ADM goes. CHS goes considerably farther due to its 50% ownership of Ventura Foods, a major producer of margarine shortening, cooking oil, salad dressing, mayonnaise and sauces.

In addition to its various brands, Ventura Foods also produces private label brands for other companies. It also is a major supplier to the food service industry in the forms of large containers and portion cups. Ventura Foods has 12 manufacturing plants, one of which is in Albert Lea, at the former Miami Margarine plant.

The plant produces about one million pounds per day of margarine and shortening for the private label and food service industries. Could a salad dressing or sauce company in Minnesota carve out a niche as a local provider? This is a question that a food industry analyst would have to answer. However, with the major brand owners already so well established it would require offering the consumer something special aside from the fact that the product was made in Minnesota.

Biodiesel

With petroleum selling for around $65 per barrel and with Minnesota law mandating a minimum incorporation of 2% of biodiesel in all diesel fuel sold at the pump in the state, it appears that the time for biodiesel has arrived. Biodiesel is the name given to diesel fuel made from vegetable oil or animal fats through a process called transesterification. The oil or fat is combined with methyl alcohol and a catalyst to form biodiesel and glycerin.

Biodiesel has been used for a number of years as a lubricity agent to replace the lubricity in diesel fuel that is lost as sulfur content is reduced in order to comply with environmental laws. Biodiesel also is has a higher cetane rating than petroleum diesel (cetane is to diesel fuel as octane is to gasoline) and it produces lower levels of sulfur and particulate emissions.

For lubricity enhancement purposes, only a very low blend of biodiesel is needed-in the range of 1 part in 400-1,000. In order to play a significant role as a fuel, biodiesel will have to be blended in much higher levels than this. A blend of 20% biodiesel, called B20, has been promoted as a means to replace significant amounts of petroleum diesel. To date B20 has been used primarily by government fleets in order to meet Alternative Fuel Vehicle requirements mandated by the Energy Policy Act of 1992 (EPAct). In order to qualify, a vehicle must use B20 or higher.

Many municipalities and school districts also use B20 (or higher) in their bus fleets, primarily to be a "good neighbor" by reducing sulfur and particulate emissions. State mandates could become a new factor in biodiesel consumption. These mandates require that diesel fuel sold within the state contain a certain percent of biodiesel. Minnesota has led the way here. A law passed in 2002 took effect on September 29, 2005, requiring that all diesel fuel sold at the pump within Minnesota contains a minimum 2% blend of biodiesel.

The start-ups in August of the Minnesota Soybean Processors (MnSP) biodiesel plant at Brewster, with capacity of 30 million gallons/year, and the SoyMor biodiesel plant at Glenville, with capacity of 25 million gallons/year, pushed Minnesota's biodiesel production well above the 8 million gallons/year production threshold within the state that the law required in order for the mandate to become effective.

According to the National Biodiesel Board (NBB), Minnesota is the only state that currently requires biodiesel content for all diesel fuel sold at the pump. Several other states do require state agencies to utilise biodiesel-blended fuels. For example, this summer Illinois passed a law that will require 2% biodiesel incorporation in July 2006 in vehicles operated by state and local governments, school districts, universities, community colleges and mass transit agencies. How many states will follow Minnesota's lead in mandating biodiesel blends in diesel fuel sold at the pump remains to be seen.

Even prior to the recent run-up in the price of petroleum, biodiesel had become more competitive with petroleum diesel by virtue of a federal tax credit for fuel blenders that is worth $1 for each gallon of "first use" agri-biodiesel they blend and 50¢ for each gallon of biodiesel made from waste grease. This tax credit was originally established as part of the American JOBS Creation Act of 2004, which had an expiration date of 2006. This was extended through 2008 by the new energy bill that President Bush signed into law on August 8, 2005. Another tax incentive for biodiesel production has been the Commodity Credit Corporation Bioenergy Program, which took effect on October 31, 2000.

The CCC Bioenergy Program called for $300 million of cash payments to companies that increase their purchases of bioenergy feedstocks. However, the importance of this program has waned with diminished funding of only $25 million for the federal fiscal year which ends September 30, 2006, at which time the program is scheduled to end. State mandates and tax credits certainly have given a boost to the infant biodiesel industry, but what will be more important to the development of the industry will be the forces of the marketplace.

When the 2000 study was conducted the price of petroleum was about $35 per barrel. Today it is around $65 per barrel. This brings the price of petroleum and the price of soybean oil much closer together than they were then. At that time soybean oil was at a very low price-only about 16¢ per pound, which equals $50 per barrel (42 gallons). Today soybean oil is worth about 22¢ per pound. This is also the average price of crude soybean oil, f.o.b. Decatur IL, over the past ten years. This equals $69 per barrel. So the price of soybean oil and petroleum are nearly identical today. What the price relationship will be in the future is anyone's guess but most observers believe that high petroleum prices will be with us indefinitely.

A good number of biodiesel plants either have been built or have been announced as planned. Not all of those that have been announced as planned will be built, but even if not, one has to wonder if biodiesel production capacity is running too far ahead of demand. Minnesota is an example. There is now production capacity operating within the state of about 60 million gallons/year. The 2% biodiesel blend mandate that took effect on September 29 will require 16-18 million gallons per year or only 27-30% of active production capacity.

The amount of B20 consumption by state and local fleets will not add significantly to this amount. However, if biodiesel blends become and remain lower in price than neat petroleum diesel, demand could increase significantly. It is too soon to tell if this is likely to happen. The original Minnesota legislation called for the mandated blend to increase to 5% at some future date but it was not included in the final version of the law. If this were to be reinstated the demand for biodiesel in Minnesota would jump to 40-45 million gallons/year-still only 75% of current capacity, but a big step toward bringing production capacity and demand into balance.

With biodiesel now being pricecompetitive with petroleum diesel, it should be politically feasible to increase the mandated blend to 5% in the near future. Better yet, market forces themselves could be the impetus to drive biodiesel blend rates beyond the required 2%. Of course there is nothing restricting Minnesota-produced biodiesel from being sold outside the state. However, there is plenty of biodiesel production capacity either already in place or being planned for the rest of the country.

In June Cargill announced plans to build its first biodiesel plant in the United States, a 37.5 million gallon/year plant at its soybean processing plant in Iowa Falls, IA. Also in Iowa, Western Iowa Energy has a 30 million gallon/year biodiesel plant under construction at Wall Lake. Start-up is planned for the first quarter of 2006. This plant will use soybean oil and animal fat as its feedstocks, with a 70/30 ratio anticipated (70% soybean oil). Four currently active smaller biodiesel plants in Iowa produce another 30 million gallons/year, led by West Central Soy at Ralston, IA, which opened a 12 million gallon/year plant in late 2003.

With the start-up of the Cargill and WIE plants next year, Iowa will displace Minnesota as the leading biodiesel producing state in the country, with nearly 100 million gallons/year of capacity. In April Mid-America Biofuels, a joint venture among ADM and three partners, announced plans to build a 30 million gallon/year biodiesel plant in Mexico, MO, where ADM has a soybean processing plant. While this will be ADM's first biodiesel plant in the United States, they already have an extensive biodiesel business in Europe. And on October 5 ADM announced plans to build a 50 million gallon/year biodiesel plant at its canola processing plant in Velva, ND, using canola oil as the feedstock.

History has shown that once ADM and Cargill get into a processing business, they expand their capacity quickly in order to achieve a dominant place in the industry. This happened in the ethanol industry and in the vitamin E industry, in the latter case to the extend that the market collapsed. It also stands to reason that a processor/biodiesel producer will have an advantage over a stand-alone biodiesel producer similar to the advantage that the processor has over the stand alone refiner.

Today almost all soybean oil is refined by the processors, with the processor/refiner having replaced the stand alone refiners-companies like Procter & Gamble, Kraft and Hunt-Wesson-who no longer do their own refining, but have it done for them by the processor/refiners, while they concentrate on marketing and distribution of their branded products. Looking at the nation as a whole, NBB lists 45 currently operating biodiesel plants and pegs current operating capacity at dedicated biodiesel plants at 180 million gallons/year. Capacity at oleo chemical plants is pegged at 110 million gallons/year.

NBB also lists 54 proposed dedicated biodiesel plants that are not yet operating. These plants can be in anywhere from the equity-drive to the under-construction stage in their development. If all these plants are built, NBB estimates that the total capacity of dedicated biodiesel plants would reach 570 million gallons/year, not including the recently announced 50 million gallon/year ADM plant at Velva, ND. NBB estimates that by May 2006 there will be another 100 million gallons/year of capacity on line at dedicated biodiesel plants. This would push capacity to 280 million gallons/year at dedicated plants, aside from the potential capacity at oleo chemical plants.

NBB projects consumption in 2005 to have doubled from 2004, to 50 million gallons. While this is impressive growth, it represents only 28% of currently active dedicated biodiesel capacity. That is to say, the supply/demand situation for the nation as a whole is virtually identical to that for Minnesota. If biodiesel consumption were to double again in 2006, to 100 million gallons/year, this would be 36% of the dedicated production capacity that NBB projects to be on stream by May, 2006. Such great underutilisation of production capacity cannot bode well for profitability within the industry.

One biodiesel producer believes that total U.S. consumption of biodiesel eventually will reach 1 billion gallons/year. This would represent about 2% of total diesel consumption in the country. But that producer believes that it will be several years before consumption reaches this point. In the meantime, it appears that biodiesel production capacity will significantly outpace consumption for at least the next few years. Eventually market forces will bring supply and demand into closer balance, but it could require some "shakeout" in the industry for this to happen if capacity continues to significantly outpace demand.

If one were to look only at the supply situation with soybean oil, the United States exported 650,000 MT of soybean oil in the year that ended September 30, 2005. In most cases, the export market is a residual market for U.S. soybean processors. If all this oil were redirected to biodiesel production it would produce about 195 million gallons of biodiesel..far short of the amount required. Could more soybean processing plants be built to produce the additional soybean oil that could be used to make up this shortfall? A soybean processing plant produces four times as much meal as it does oil and the market for soybean meal in the United States is already saturated to the point where the United States exports 4-6 million MT per year even though in most cases the export market is the residual market for soybean meal.

If enough additional soybean processing capacity were built to supply, say, an additional 250 million gallons/year of biodiesel-over and above the 195 million gallons that could be made from the oil that was exported this year, and the amount of "domestic" market oil that is already being used to produce 50 million gallons/year of biodiesel..to bring total biodiesel production from soybean oil up to 500 million gallons/year-it would require 4.5 million MT/year of additional soybean processing capacity. This equals five more crush plants the size of the new ones at Fairmont and Brewster. These plants would produce 3.4 million MT/year of soybean meal.

More soybean meal from the United States is not needed in the world market. The peak years of U.S. soybean meal exports are behind us. We exported 8.5 million MT in 1997/98 and 5.7 million MT in 2004/05. Argentina (20.6 million MT) and Brazil (15 million MT) have largely displaced the United States as the suppliers to the world soybean meal export/import market of 47 million/MT/year. And additional crushing capacity continues to be built overseas in countries that formerly were large importers of soybean meal.

The United States consumed 36.2 million MT of soybean meal in 2004/05. Consumption over the previous five years averaged 34.8 million MT. Soybean meal consumption in the United States is roughly paralleling population growth, which is about 1% per year. At this rate it would be nine years before the domestic market can absorb another 3.4 million MT of soybean meal. However, during that period it is likely that soybean meal exports from the United States will continue to decline, with this meal being re-directed to meet the slowly increasing domestic demand.

Whether additional soybean processing capacity will be built to provide oil to the biodiesel industry will depend on how fast the biodiesel industry grows. If it grows very slowly to 500 million gallons/year..and then to 1 billion gallons/year..it is possible that some additional soybean processing capacity could be built to fill some of this need. But even in this case it is not likely, due to the situation with the meal. If the biodiesel industry grows quickly, feedstock other than soybean oil will surely be needed.

Soybean oil accounts for 81% of vegetable oil production in the United States. The second largest is corn oil, which accounts for 10%. It is too valuable to be used for biodiesel production. The same is true of the #3 oil, cottonseed (4%) and the #5 oil, sunflower (1%-144,000 MT). Production of canola oil is too small to make a large contribution to biodiesel feedstock. Nearby Canada is a large producer of canola oil-1.4 million MT in 2005-and a net exporter of 875,000 MT, with the United States accounting for over 500,000 MT. How much of this might be diverted from food use to fuel use is hard to say.

There will be competition from Europe for Canadian canola oil as a biodiesel feedstock. Since most Canadian canola is GM varieties, Europe has not been a destination for Canadian canola oil as a food source. However, this is not an issue when canola oil is used as a biodiesel feedstock. German-based oilseed analyst Oil World recently reported that Europe will import 75,000 MT of Canadian canola oil in the 2005/2006 marketing year, of which 50,000 MT already has been purchased and 25,000 MT shipped.

Animal fats and recycled vegetable oils and greases will play an increasingly large role if the biodiesel industry expands fairly quickly. It already has been mentioned the WIE plant in Wall Lake, IA anticipates using 30% animal fat in its feedstock mix. The FUMPA plant in Redwood Falls includes animal fats in its feedstock mix. According to the National Renders Association, the United States produces about 3 million MT of inedible tallow and greases annually, 900,000 MT of edible tallow, 115,000 MT of lard, and 400,000 MT of poultry fat. Of these amounts, a combined 1.5 million MT, or about a third, are exported.

Assuming a biodiesel yield equal to 90% of the yield from soybean oil, these exported fats and greases could produce around 400 million gallons of biodiesel. The 3.0 million MT/year of fats and greases that are consumed domestically could produce another 800 million gallons of biodiesel. So there are enough biodiesel feedstocks to produce 1 billion gallons/year of biodiesel in the United States, if market forces are able to divert enough oils, fats and greases that are currently exported or consumed by the domestic feed industry.

How much waste grease would be available for biodiesel production is anyone's guess. Domestic consumption of all fats and oils in the United States is about 14.5 million MT. Some of this is consumed in home use in the forms of butter, cooking oil, margarine, salad dressing, sauces, soaps and detergents, and cosmetics. It is unlikely that spent home-use fats and oils would be recycled. Large quantities of fats and oils are consumed as a food ingredient. This includes oil consumed as fry oil by the baking, snack food, fast food and restaurant industries.

Frito Lay is the largest user of vegetable oil in the world. The fast food and restaurant industry consumes large quantities in the preparation of French fries. Spent oil from food companies and the fast food and restaurant industries goes mainly to the feed industry. As in the case of rendered fats and greases, the biodiesel industry could compete with the feed industry for this tonnage.

Current prices for rendered products range from 16¢/pound for yellow grease to 20¢/pound for tallow, with poultry fat in between. This makes rendered products about 10-20% cheaper than soybean oil but the biodiesel yield is about 10% less. The yield on waste grease is worse yet, perhaps 20% worse than from soybean oil, and the blender's credit is only 50¢/gallon on waste grease rather than $1/gallon for first use agri-biodiesel. One biodiesel producer stated that they cannot currently make the economics work using waste grease as the feedstock, citing high collection costs as a factor in addition to the poorer yield and lower blender's credit.

One biodiesel producer thinks that eventually 15% biodiesel incorporation is possible. Based on national diesel fuel consumption of 50 billion gallons/year, this would be 7.5 billion gallons of biodiesel. This would require 25 million MT of feedstock. Total fats, oils and rendered fats and greases production in the United States is only 20 million MT/year, most of which is for food use. So it is hard to imagine where 25 million MT of feedstock would come from. This illustrates the point that although biodiesel has a role to play in the fuel industry, it is not going to make a huge dent in petroleum diesel consumption.

Uses of Soybeans

Soybean oil is used in non-food products such as ink, paints, adhesives, waxes, solvents, cleansers and lubricants. In these applications it is seen as more environmentally friendly than petroleum-based counterparts in addition to being renewable.

Linseed oil was the most common "vehicle" for paints prior to the introduction of latex paints many years ago. It continues to be used in paints and stains today but to a much lesser extent than in its hey day. This is reflected in the Minnesota flax crop being only a fraction of it what it was back then-a mere 1295 MT in 2004 compared with the record crop of 485,000 MT in 1948. Soy oil-based ink has become quite popular.

Many companies and organisations that use it make mention of the fact in their printed material to demonstrate their commitment to using a renewable, non-petroleum-based product that also is environmentally friendly and provides a more healthful work environment in print shops. Mallard Ink of St. Anthony produces a small amount of soy oil-based but points out that soy oil-based ink has some functional characteristics that make it less desirable for some printing applications than linseed oil-based ink, which accounts for most of Mallard Ink's production.

At the time of the 2000 study a promising new composite made from soy flour and recycled newsprint was being introduced by Phenix Biocomposites of Mankato. The company had been producing its granite-like composite board, called Environ®, at a small plant in St. Peter for a few years and had recently commissioned a much larger plant in Mankato.

Eventually a new ownership group emerged and changed the company name to Environ Biocomposites. Under its new ownership the company has continued to struggle, particularly with its namesake product. It has produced a great deal more of a cheaper, particle-board-like product made from wheat straw called BIOFIBER™, which also had become its primary product back in 2000. Since 2000 it has introduced a new product called Dakota Burl™, a decorative product made from sunflower hulls.

Recently Environ Biocomposites was sold to an investor in California. The closing is expected to occur in November of this year. Hopefully the new ownership will be able to solve the company's financial and marketing problems that have plagued it since the Mankato plant was built in 1999.

A new industrial use of soybean oil is the production of polyol, which can be used as a placement for petroleum-based polyol in the production of polyurethane. SDSP produces polyol under the trade name SoyOl™which goes into products as diverse as carpet backing and automobile dashboards. Dow Chemical has been a leader in the development of soy-based polyols. At this point this is a small business but the growth potential is good.

As with other products made from soybean oil, a processor should have an advantage over a manufacturer who must buy his raw material from a processor and transport it to the production site. This topic was addressed in the previous study in some detail in an attempt to quantify the disadvantage to the crusher posed by Minnesota soybeans, which are historically lower in protein content than soybeans grown farther south.

The calculation worked out to be 11¢ per bushel, which is a very significant amount to a crusher. Since then the United Soybean Board has launched a program to improve the protein and oil content of northern soybeans. The target is 35% protein and 19% oil. The seed companies have gotten on board and are promoting their varieties that have been shown to give these yields and some of the processors will pay a premium for beans that have come from the recognised varieties. Beans containing 35% protein, 19% oil and 13% moisture would produce hipro meal with 48.1% protein content, assuming an oil extraction rate of 95% and meal moisture of 12% moisture.

Several processors now have gone to "component pricing" whereby they pay a premium for protein and/or oil content over certain levels or pay a premium for certain varieties of soybeans which have been shown to achieve the 35% protein/19% oil target.

REFERENCES

Labys, W. C. 1977. "Multicommodity Substitution Patterns in the International Fats and Oils Market." *European Review of Agricultural Economics*, Vol. 4, No. 1, pp. 75-84.

Landell Mills Commodities Studies, Inc. 1991. *Oils and Oilseeds: The Cost Competitiveness of Major Producing and Processing Countries*. New York, NY.

Meilke, K.D. and M. Jay. 1997. *Zero-for-Zero and the Canadian Oilseed Complex*. A report prepared for Agriculture and Agri-Food Canada, Ottawa, Ontario.

Meilke, K.D. and M. Swidinsky. 1998. *An Evaluation of Oilseed Trade Liberalization*. A report prepared for Agriculture and Agri-Food Canada, Ottawa, Ontario.

OECD. 1994. *The World Oilseed Market: Policy Impacts and Market Outlook*. Paris, France.

9

Impact of Trade Liberalisation on Oilseed Sector

The oilseed complex in India is undergoing visible changes in the new environment of liberalised trade. Consumption patterns are changing, as consumers are beginning to accept oils other than those consumed traditionally. Changes in cropping patterns have also taken place with the help of technology missions and price support. Although India ranks among the largest producers of oilseeds in the world such as USA, China and Brazil, its productivity is quite low. The low and fluctuating yields are primarily due to a large part of the cultivation being on marginal lands lacking irrigation and with low levels of input usage.

Three oilseeds: groundnut, soybean and rapeseed/mustard, together account for over 80 per cent of aggregate cultivated oilseeds output. The processing technology is a mix of the traditional and modern ranging from household crushing to expander/extruder technology in medium-scale factories. There has been a large growth in processing capacity partly due to de licensing of the vegetable oil industry in 1990-91 and partly due to tax incentives. Processing of oilseeds such as groundnuts and rapeseed however remained inefficient due to the existence of small-scale industry reservation policy that prevented efficient operations.

The edible oils/ oilseeds sector currently faces several challenges. Oilseed cultivation is becoming increasingly unattractive due to low and unstable yields. The technology mission on oilseeds had only limited success. Decreasing price of edible oils due to trade liberalisation may result in low prices for oilseeds resulting in poor supply response. High import tariffs and nontariff barriers such as sanitary and phyto sanitary (SPS) restrictions have made oilseed imports unattractive. Low domestic output of raw material combined with restricted import of oilseeds can lead to a high degree of under utilisation of processing capacity.

The government's help to oilseed growers has been in the form of providing Minimum Support Prices (MSP)) through its stocking policy and by imposing customs duties on imports of edible oils and oilseeds. MSP policy does not appear to have worked as well in the case of oilseeds as it has been in the case of wheat and rice. Import tariffs on edible oils tend to impose a large burden on consumers and help processors more than oilseed farmers.

Government intervention needs to balance the interests of different stakeholders in the oilseed complex. Protecting the oilseed growers could make oil and meal products internationally uncompetitive. Low priced imported oils benefit the consumers but tend to reduce the margins on domestic oils affecting processors and oilseed farmers adversely. Thus, with trade liberalisation several issues arise, including the choice between protecting the seed sector as opposed to the processing sector.

Both consumption patterns and cropping patterns are likely to be influenced by the choice of customs duties and the price differences maintained among various oils. Imported oils account for close to fifty percent of the total edible oil consumption. For example, soy oil is competitive due to its low price and the low duty it faces. Refined Palm oil is also competitive due to its low price, even though it bears a high duty. This could affect the prospects for other oils such as sunflower and rapeseed oils both of them being priced relatively high and bearing higher duties. The impact of such changes can have varying effects on consumers from different regions with varying tastes.

TRENDS IN OILSEEDS PRODUCTION

India is one of the largest producers of oilseeds in the world. Its share in world production is as high as 27 percent for groundnut, 23 percent for sesame, 16 percent for rapeseed and 66 percent for castor seed. Groundnuts in shell form 11% of world exports and sesame forms 17%. India's productivity is however quite low, around fifty percent of the world average and even less in the case of soybean (Table 1).

Table 1: Productivity of oilseeds: India and the world (tonnes/ hectare)

Oilseed	*India*	*World Average*	*Highest*
Soybean	0.85	2.29	3.28(EU-15)
Cottonseed	0.59	1.06	2.07(Australia)
Groundnut	0.59	1.02	2.13(China)
Sunflower	0.62	1.18	1.73(EU-15)
Rapeseed/Mustard	0.75	1.49	2.96(EU-15)

The comparatively lower yields are mainly due to the fact that the quality of the seed varieties is generally poor and oilseeds crops in India are mostly cultivated in unirrigated areas. Less than 25% of cropped area is under irrigation. For the same reason yields are more variable due to weather fluctuations. Other reasons include disease and pest damage, vulnerability to drought, poor dry farming practices, low access to inputs and poor soils.

A large number of oilseed crops are grown in different regions under different agro climatic conditions. These crops are among the most widespread in small farm systems throughout India. Groundnut, rapeseed-mustard, sesame and safflower are the traditionally cultivated oilseeds. In the recent years however soybean and sunflower have gained importance.

Starting from the late seventies India depended heavily on imports to meet its edible oil demand. This prompted the setting up of the Technology Mission on Oilseeds in 1986, which gave a thrust to the production of oilseeds. Production of oilseeds went up from around 11 million tones to around 25 million tonnes towards the late nineties. There was almost a 2 percent growth in yields over the last fifteen years. Most of this growth came from soybeans, rapeseed-mustard and groundnuts.

In the recent years however there is stagnation in both area and production. The import substitution policies followed by the government until recently have provided incentives to farmers by raising oilseed prices relative to competing crops. Although, oilseeds have been opened up for trade recently the tariff rate of 30% appears to be high making imports unattractive. Other restrictions such as the SPS barriers also prevent imports. For example, the requirement that imported soybeans be split is a deal breaker, since no exporter is set up to do this. If oilseed imports have to be GMO free, sunflower seed is the most likely choice since it can be certified GMO free if imported from Australia.

In the case of soybeans and rapeseed there is no origin that can certify a GM free shipment. In the simulation exercises using our model, "trade liberalisation" scenarios assume away the existence of non-tariff barriers. Such a scenario would obviously affect the production structure of oilseeds depending on the comparative advantages in domestic production for given world market situations.

Groundnuts

India ranks second in the world in groundnut production.

The three southern states of Andhra Pradesh, Tamil Nadu, Karnataka and the western state of Gujarat together account for close to 80% of the annual output in India. About 70-75% of the crop is Kharif, grown during summer rainy season (planted during May-July and harvested in September-mid December). In the Rabi (winter) season planting is during mid September to November and harvesting during March and April. Crop failures occur periodically due to inadequate or excessive rain or unfavorable rainfall distribution.

Groundnut ranks among oilseeds with high oil recovery (40%). Around 40 to 50% of the output is used in oil production the rest being used as seed and feed. Regional trends in groundnut production indicate that the recent increase in groundnut yields has mainly occurred in Tamil Nadu due to increased irrigation. Although Tamil Nadu accounts for 12% of the total area under groundnuts, it contributes to 22% of the total production.

Rapeseed-Mustard

India ranks fourth (after China, EU and Canada) in the world in the production of Rapeseed. In the case of Rapeseed mustard there is 33% oil recovery. Almost the entire output (97%) is used in oil production is used in oil production with only 3% used for seed and feed purposes. Almost 40 percent of the rapeseed output comes from the state of Rajasthan (Table 2). Other major states include Uttar Pradesh (18%), Madhya Pradesh (10%) and Haryana (11%). Yield improvements have taken place in all the major states although Haryana shows the maximum growth in yields in the last two decades. However, variability in yields is also higher for this state. Area under rapeseed-mustard crop gained substantially from the expansion of irrigation. Import of rapeseed/mustard oil peaked during the late 1980s with imports of around 0.19 million tonnes in 1988-89. These imports reduced drastically with the increase in domestic production of rapeseed/mustard and also due higher world prices of rape oil in recent years as compared to the availability of cheaper imported oils like palm.

Table 2: Regional pattern of oilseed production: 1998-99 to 2000-01

Oilseed	*Major producers*	*Percent of total Area*	*Production*
Groundnut	Andhra Pradesh	27	25
	Gujarat	26	19
	Karnataka	17	14
	Tamil Nadu	12	22
Soybean	Madhya Pradesh	70	66
	Maharashtra	18	22
	Rajasthan	10	10
Rapeseed-Mustard	Rajasthan	45	40
	Uttar Pradesh	17	18
	Madhya Pradesh	10	10
	Haryana	8	11

Soybeans

Soybean is predominantly grown in the states of Madhya Pradesh, Maharashtra and Rajasthan. Madhya Pradesh grows around 66 percent of the country's soybean output occupying around 70% of the total area under the crop (Table 2). It is generally grown as a rainy season crop under rainfed situation (sown in June and July and harvested in November and December). Since soybean is cultivated as a rainfed crop, yields are highly variable. Higher and more stable yields are possible through assured minimum irrigation and drought resistance brought about by genetic improvement. Weeds and pests also contribute to yield losses. The crop has exhibited a vast potential as 'Kharif' (monsoon season) crop mainly in the Central India, and is extending its coverage in the Southern parts of the country.

There has been a rapid growth in soybean cultivation over the last two decades placing India on the world map of soybean. Yield growth was the highest in the case of Maharashtra, growing from close to 600 kg per hectare in 1989-90 to 1100 kg per hectare in 2000-01. India ranks fifth in the world in area and production after USA, Brazil, China and Argentina. In recent years it has become the second most produced oilseed in India.

The phenomenal growth in cultivation of soybean can be attributed to the concurrent development of the soy-oil/meal industry, which provided remunerative market to the growers. Soybean processing capacity grew very fast with the establishment of the first solvent extraction plant for soybean in 1972 at Indore (Madhya Pradesh). Roughly 85% of the output goes into oil extraction, the rest being used for food feed and seed purposes. Recovery of oil is about 18% and that of soy meal is about 82%.

Since the domestic consumption of soy meal is limited most of it is exported in the form of deoiled cake. Soybean meal produced in the relatively modern solvent-extraction plants was of acceptable quality for export and found a ready market. There was no Small Scale Industry reservation policy for soy meal as opposed to other meals produced in small-scale crush facilities. The export earnings have been quite significant. India earned Rs. 19 billion exporting 2.5 million tones of soy meal in 2001-2002. South Korea was the largest importer of Indian soy meal, followed by Indonesia and Japan.

OILSEED PROCESSING SECTOR

The edible oil processing sector in India is characterised by large unutilised capacities. Capacity utilisation is higher in the refining sector mainly due to import of oil in crude form. Inefficiencies in the processing sector are mainly due to highly fragmented nature of the industry.

The small-scale industry (SSI) reservations applying on groundnut and rapeseed expellers prevent vertical integration of expeller and solvent extraction units like in other countries. This limits the ability of the processing units in reaping economies of scale. There is also the problem of excess capacity creation and production inefficiency due to under underutilised capacity. Excess capacity in soybean crushing for example, could be attributed the "replenishment" scheme that gave import licenses to soy meal exporters to import any goods of equivalent value, and also due to expectations of rapid growth in soybean production.

The reason for overinvestment was due to tax holidays and other incentives provided to build plants (e.g. for development of backward areas). These incentives led to a uniform geographic spread of capacity. However it did not match up with the availability of raw material in the different regions. Under utilisation of capacities is therefore mostly due to lack of raw materials and the impossibility of import of oilseeds due to tariff and non-tariff barriers. The processing sector is also characterised by low profit margins due to stiff competition including from imports. The SSI policies apply mainly to the expeller units and not the solvent extraction units. The traditional expeller units (Ghanis) operate at just 10 percent capacity utilisation whereas the more modern ones operate at 30 percent. In the north there is a strong preference among consumers for oil from Ghanis due to the natural flavor. The share of solvent extracted and refined oils in the domestic edible oil supply is increasing however.

Experts in the industry believe that solvent extraction and refining units are undergoing a phase of consolidation and becoming more concentrated. Currently around 25% of the turnover is in the hands of 10-12 players out of a total of 600 players. By 2010 this expected to grow to 50%. There is however a limit to the extent of concentration because of the wide variety of oils/oilseeds and regional difference in tastes across a large country. Solvent extraction is a method of extraction of oil using a chemical solvent which is ideal for oilseeds with low oil content, e.g. soybean and cotton seed. This technology is however being used for rapeseed-mustard and groundnut in order to circumvent the SSI policy restriction. The expander-solvent plants have been more successful with rapeseed compared to groundnuts.

EDIBLE OIL DEMAND IN INDIA

Edible oil consumption in India has been growing steadily over the years. Aggregate consumption increased from around six million tonnes in the early nineties to around than 11 million tonnes in recent years. However, per capita consumption of fats and oils in India is far below the world average (Table 3).

Table 3: Comparing India's consumption of oils and fats with major countries (Kgs percapita per annum)

	1997	*1998*	*1999*	*2000*	*2001*	*2002*
Eu-15	43.6	44.8	45.8	47.1	49	49.7
Senegal	20.6	20.6	20.3	18.8	20.5	21.5
Peru	14.4	14.1	14.1	13.6	13.6	14.1
Ex-USSR	12.8	13.2	14.1	14.9	16.3	17
Nigeria	12.1	12.2	12.6	13	13.1	13.3
Chile	24.3	25	25.1	27.8	27.8	27.6
Canada	44.2	45.6	45.5	45.4	45	44.3
USA	46.5	46.8	49.8	50.3	51.7	51.3
Mexico	22.4	23.3	23.8	23.9	24.4	25.2
Argentina	23.6	23.7	23.3	23.1	21.3	19.8
Brazil	22.9	24.5	23.8	25.8	24.7	24.4
China PR	12	12.5	12.9	14	15	15.9
India	9.9	10.1	11.2	11.6	11.8	11.8
Indonesia	15.7	15.3	15.2	15.6	16	16.1

Japan	21.3	20.8	21.2	21.6	21.8	22
Korea South	18.9	17.1	19.7	20.6	21.1	21.8
Pakistan	18.7	19.1	19.3	19.7	19.5	19.6
Philippines	7.5	6.7	7.8	7.6	7.6	7.8
Turkey	26.7	27.3	26.5	26.7	24.5	26
Australia	33.8	32	32.6	32.5	31.9	31.7
New Zealand	27.3	29	29.3	31.9	34.1	37.5
World Average	17.3	17.47	18.12	18.69	19.23	19.52

Groundnut, Rapeseed/mustard and Soy oil together form around 60% of edible oils consumed in the country. Trends at the disaggregated level indicate that palm and soy oil, the two major imported oils, contributed the maximum to the growth in consumption.

Increasing per capita incomes and availability of cheaper imports due to trade liberalisation are some of the important factors behind this increase in demand. India's imports of palm and soy oil form 17% and 6% respectively of world imports of palm and soy oil. Consumption preferences for different oils vary across different regions in India and also between rural and urban areas. Groundnut and mustard oil together account for 59% and 67% of total edible oil consumption in urban and rural areas respectively. Consumer preferences are influenced by the crops grown in their regions. For example, in most of the southern and western states there is a strong preference for groundnut oil, whereas in the east and the north it is mustard oil. The northern states also consume Vanaspati, a partially hydrogenated edible oil mixture.

Oils that are generally not consumed directly for taste reasons (e.g. soybean, rice bran and cotton seed) are used in the manufacture of vanaspati. Newer oils such as soy oil and palm oil are finding their way into direct consumption mainly because of the cheaper price even with high tariffs. Market shares of raw oil, refined oil and vanaspati are respectively 42%, 43% and 13% respectively. About 60-70% of groundnut and mustard seeds are used to make non-refined or filtered oils as consumers have a liking for their distinct flavor.

Non-refined oils are also mostly bought in loose form (unpackaged). However, consumers, at least in the urban areas are beginning to shift to packaged/branded oil due to growing heath consciousness. The available estimates of demand show that edible oil consumption is highly price and income elastic. It is to be noted that aggregate national level consumption of estimated based on NSS data are underestimates compared to the supply and use data published by sources such as Oil World, Solvent Extractors Association and USDA. One of the reasons for this is that some of the intermediate consumption of edible oils is not reflected in the household survey data. For example, a large part of palm oil goes into the manufacture of 'Vanaspati', which is mostly used in commercial establishments such as restaurants, bakeries, sweet shops etc. The indirect consumption of vanaspati through foods purchased at these outlets is not however reflected in the NSS data. This is reflected in the fact that Vanaspati consumption is confined mainly to four states, Haryana, Punjab, Uttar Pradesh and Himachal Pradesh according to the NSS data.

Oilseed Crush Demand

Total demand for seed is the sum of crush demand by processors and other demand (food, feed etc). The latter is specified exogenously and the former is a function of crushing margin and income. Crushing margin depends on the prices of oil, meal and seed and the extraction rates of oil and meal from the seed. It is defined as

$$cm = a\ p^{oil} + b\ p^{meal} - p^{seed}$$

where 'a' and 'b' denote the technical extraction rates (amount of oil/meal by weight per unit weight of seed) of oil and meal respectively.

Oilseed crush demand is motivated by the size of the crush margin relative to the cost of crushing. Since processing of oilseeds is not modeled here it is implicitly assumed that average costs are constant. This specification is appropriate only for a short run analysis and the changes in crush demand would indicate movement along the demand curve and not shifts in the curve. Supply: Total supply of each oilseed is taken to be the aggregate of oilseed supply from all the producing states plus imports. Supply of oilseed in each of the producing states is a function of its own price, among other variables.

Oilseed Imports

Oilseed imports are positive when domestic demand exceeds domestic supply at the import parity price. Market equilibrium price in such a case is equal to the import parity price. However, if at the import parity price there is excess supply then imports are zero and equilibrium price would be lower than import parity price. Thus, equilibrium domestic seed price is always lower than or equal to the import parity price.

Note that for domestically produced oilseeds the wholesale price is obtained as price at farm level plus wholesaler and marketing margins. In general, marketing margins account for the price difference between the wholesale market and the farm gate and consist of transport costs, taxes, insurance, interest charges, bagging, packing and handling charges, grading and storage. In the case of commodities that require processing, processing costs are also included. For imported oilseeds, import parity price (at domestic wholesale market level) is obtained by adding external trade and transport margins to the border price. External trade margins include foreign exchange brokerage, import registration fees, domestic freight, port charges, import taxes, VAT and other domestic taxes.

Price Support to Oilseed Farmers

Equilibrium price of seed is influenced by government's price support policy. The government can support farmers' price at a particular level (MSP) through alternative mechanisms. Through tariffs on import of edible oils, through tariffs on import of oilseeds or through subsidies similar to the marketing loan program or loan deficiency payment scheme in the USA. In the first alternative consumers subsidise farmers and processors. In the second alternative subsidy to farmers is borne by both processors and consumers. In the last alternative the cost of subsidy to farmers is borne by the government.

Demand for each of the oils at the all India level is obtained as the aggregate of state-wise demands. In each state, demand for oil is a function of own and substitute oil prices and income. Since India is a net importer of edible oils, supply consists of domestic production and imports. Domestic production of oil is determined from the crush demand for seeds, using the technical extraction coefficients.

As in the case of seeds, net imports are equal to the excess demand (demand less domestic supply) for oil at the import parity price and in general domestic market equilibrium price is less than or equal to the import parity price. In the model import parity prices are made responsive to the magnitude of imports ('large country' assumption). For example, in the case of palm oil the volume of imports by India is a substantial proportion of world trade (17%). World prices in this case may respond to the magnitude of imports by India and cannot be taken as given.

India is a net exporter of oil meal and hence supply of meal oil consists of only domestically produced meal. As in the case of oil, meal output is determined from the crush demand for seeds using the technical extraction coefficients. Total demand for meal consists of domestic consumption and exports since oil meal is a net exported commodity. Export of meal is equal to the excess supply of meal at the export parity price.

POLICY TRENDS

Until recently imports of edible oils were canalised through the State Trading Corporation. Restriction on imports and stagnant domestic production implied that consumers faced high prices for oils. In order to reduce dependence on imports and increase domestic production the technology mission on oilseeds was initiated in 1986. Market intervention operations were also initiated by the National Dairy Development Board (NDDB) to stabilise prices. Import substitution policies helped in providing favorable price incentives with higher prices relative to competing crops. Nominal protection coefficients calculated for the major oilseeds, groundnut, rapeseed, soybean and sunflower were very high in most of the years.

Due to the market access commitments under the WTO quantitative restrictions on imports had to be removed. In 1994, edible oils were placed under Open General License (OGL) so that private traders could import edible oils freely subject to a tariff that was initially set at 65 percent. As per the agreement under the WTO India fixed its bound rate of tariffs at a high level of 300 percent for most oils, with the exception of soy oil for which the bound rate was fixed at 45 percent. Despite this low bound rate, palm oil and not soy is the most imported oil in India. This is mainly due to the price sensitivity of the Indian consumers. Not only are palm oil prices lower at the point of origin (Malaysia and Indonesia) but also the freight cost is cheaper compared to Soy oil supplied from USA or Argentina. With trade liberalisation imports rose and the prices of edible oils dropped for consumers. Currently more than 40 percent of India's edible oil demand is met through imports. This was mostly due to the gradual reduction in applied tariffs on all oils. By 1998 the tariff rates were reduced to 15 percent. In 1997 a tariff surcharge and special additional duty were added to the basic duty bringing the rate to 16.5 percent. These rates were however applied uniformly across all commodities. Of late there have been frequent adjustments to tariffs to protect domestic oilseed producers and processors and to smooth the effect of fluctuation world prices on domestic consumers. Differentiated tariffs were applied to different oils such as palm and soy.

Duty difference was also made between crude and refined imported oils. This policy helped value addition within the country and also prompted modernisation and capacity addition in the processing industry. The difference was maintained at 27.4 percent until April 2003 when it was reduced to 5 percent due to reduction in duty on refined palm oil imports. The reduction in duty on refined palm oil prevent oil prices from rising too much due to the drought conditions experienced in many parts of the country. This change has however affected adversely the processing sector. Recently the government is also taking steps to divert land areas under cultivation from wheat-rice to oilseeds through the freezing of minimum support price (MSP) for wheat and paddy. Although the government has allowed the import of oilseeds, there has been no import of oilseeds due to quarantine and other safety measures imposed by the government.

IMPACT OF TRADE LIBERALISATION

Trade liberalisation is modeled as elimination of existing import tariffs on all edible oil imports. The overall welfare impact of this change is positive. Consumers gain while the other agents lose (table 4). However, the substantial gain that accrues to consumers compensates producers, processors and government.

Table 4: Welfare impact of reducing tariffs on edible oil imports

(Rupees crores)

Change in consumer surplus	8033.39
Change in producer surplus	-2695.63
Change in processor surplus	-4863.70
Change in government surplus	-4.49
Total change in surplus	469.58

Elimination of tariffs leads to a reduction in the domestic price of all oils. It results in almost a 34 percent reduction in the average price of all oils. Price reduction is higher for imported oils. The percentage decreases in oilseed prices are much lower than those of edible oils as these are indirectly affected.

Table 5: State-wise impact on oilseed production due to trade liberalisation (Percentage changes in oilseed production due to elimination of import tariffs on edible oils)

	Groundnut	*Soy*	*Rape*	*Total*
Andhra Pradesh	0.00			0.00
Assam			0.00	0.00
Bihar			0.00	0.00
Gujarat	0.00		0.00	0.00
Haryana			0.00	0.00
Himachal Pradesh				
Karnataka	0.00			0.00
Kerala				
Madhya Pradesh	0.00	-12.28	0.00	-10.78
Maharashtra	0.00	-12.29		-9.54
Orissa				
Punjab				
Rajasthan	0.00	-12.24	0.00	-3.55
Tamil Nadu			0.00	0.00
Uttar Pradesh			0.00	0.00
West Bengal			0.00	0.00
Total of 16 states	0.00	-12.28	0.00	-5.22

Consumption/import of palm oil increases by 74% while consumption and import of other oils decrease due to cross price effects. Production of oilseeds and crushing margins also decrease as import tariff on palm oil is decreased. The decrease in production is greater in the case of soybeans. Percentage decrease in crush margin is largest for rapeseed. Although there is an overall increase in the consumption of edible oil due to decrease in import tariff on oils there is a large variation across oils in terms of percentage changes in consumption. While consumption of Soy and Palm oils increased that of the other oils decreased.

Regional impact of liberalisation is obtained mainly in terms of changes in oilseed production and producer surplus across states. Obtaining state wise changes in consumption of edible oils was not possible in the absence of state wise price elasticities of demand.

Table 6: State-wise impact on producer welfare due to trade liberalisation

	Change in Producer surplus
Andhra Pradesh	-146.78
Assam	-0.49
Bihar	-26.60
Gujarat	-373.83
Haryana	-42.28
Himachal Pradesh	-0.49
Karnataka	-200.46
Kerala	-0.49
Madhya Pradesh	-976.23
Maharashtra	-539.56
Orissa	-39.90
Punjab	-26.60
Rajasthan	-175.64
Tamil Nadu	-53.19
Uttar Pradesh	-53.19
West Bengal	-39.90
Total of 16 states	-2695.63

The impact on oilseed production due to trade liberalisation varied due to variations in production patterns as well is price elasticities of supply. As import tariffs on oils are reduced, there is a reduction in output of soybeans in Madhya Pradesh, Maharashtra and Rajasthan. The changes in producer surplus show that Gujarat, a major producer of groundnuts and Madhya Pradesh and Maharashtra (major producers of soybeans) are the top losers.

Impacts of Protecting Oilseed Growers

Three alternative ways of support to oilseed farmers' prices are considered in the model simulations. Price received by the farmers can be increased either through import tariff on edible oils or more directly tariff on import of oilseeds themselves or alternatively through government cash subsidy program.

The impact of the first alternative is obtained by comparing equilibrium oilseed prices in scenarios with and without oil tariffs. For meaningful comparison prices of oilseeds, in the other two alternatives, have to be supported at the same levels as those obtained in the case of tariffs on oil imports. That is, in the second alternative, the model would determine import tariffs on oilseeds endogenously so that prices are supported at these levels. This of course is possible only if the oilseeds are all imported.

In the third alternative the government fixes Minimum Support Prices (MSP) at levels corresponding to the prices obtained in scenario with oil tariffs. It provides a cash subsidy to the farmer equal to the difference between the equilibrium market price and MSP whenever the former goes below the latter. That is, while buyers face the market equilibrium prices of seeds, farmers receive the higher of MSP and market equilibrium price. The 'effective' price that enters the oilseed supply response function is therefore the greater of the two: MSP and the equilibrium price.

Government subsidy in this case is computed as the product of oilseed output and the difference between the 'effective' price and market equilibrium price.

Comparison of welfare impacts indicates that government cash subsidy is the most efficient alternative of protecting oilseed producers. The alternative of imposing tariff on oilseed imports turns out to be an ineffective instrument to support farmers' prices. This is because import parity prices are such that imports are unattractive for most of the oils. The alternative of government subsidy results in a lower welfare loss compared to import tariff on oils. Funding of price support in the former alternative is from the general budget as opposed to an implicit tax on edible oil consumers in the latter case. Net welfare change in fact turns out to be positive with government subsidy when domestic oilseed output is normal and international oilseed prices are lower than normal (by 10%).

The welfare impact on different agents indicates the following. Import tariff on oils benefits the processors as well as the oilseed producers. The benefits to the former are greater. These benefits however, come at the cost of consumer welfare. Price support through government cash subsidy yields positive welfare change to consumers, producers and processors. Benefits to oilseed producers are the maximum. The cost in this case is borne by the government. Government costs are higher in the case where domestic oilseed output is higher than normal.

Tariff on oil imports increases edible oil prices (average of all oils) by 51%. However, total edible oil consumption decreases only by 2% approximately. Consumption of imported oils- soy and palm decreases but that of other oils increases. Oil imports decrease by 1.765 million tones. This however is compensated by an increase in domestic production of oils by 1.214 million tones. Total domestic oilseed production increases by 7%. There is an increase in crush demand and crush margins for all seeds. Model simulations reveal that import of rapeseeds increases when tariffs are imposed on edible oil imports.

In reality, we do not see this happen due to the existence of non-tariff barriers such as sanitary and phyto sanitary restrictions. In the second alternative where producer price is supported through government cash subsidy we see that there is not much change in domestic edible oil consumption. This is due to the way the price support mechanism operates. Since cash subsidy is provided to farmers to protect their prices at a higher than the level determined by market mechanism the prices faced by processors for the seeds purchased remains the same as in the base scenario. Hence there is not much change in prices faced by domestic edible oil consumers.

Domestic production of oilseeds however increase by the same percentage as in the case of price support through tariff on oils since the level at which the prices are supported is the same in both the alternatives. However, the increase in domestic production of oilseeds leads to only a 5% decrease in oil imports, as the absence of tariffs makes imports attractive. The crush demand for seed and the domestic production of oil both increase by a lower percentage (8.5% and 6.4% respectively) compared to the first alternative.

GLOBAL SCENARIO

Oilseed production and processing occurs all over the world, but important producing and processing countries are the United States, Brazil, Argentina and China for soybeans; Indonesia and Malaysia for palm oil; and China, the European Union and Canada for rapeseed. There is a diverse array of products involved in or competing with the oilseed sector. There are seven major oilseeds: soybeans, rapeseed (canola), cottonseed, peanuts (groundnuts), sunflower seed, copra (from coconuts) and palm kernel. From the perspective of international trade, a handful of commodities and countries dominate. In fact, soy, palm and their derivatives account for 85% of all oilseed sector export volume. Today soy and palm account for 63% of world vegetable oil production, while soybean meal alone accounts for two-thirds of world protein meal production.

With the proliferation of bilateral and regional free trade agreements, tariff levels can be quite variable. With the exception of China and India, tariffs are quite modest in this sector. Complicated and less transparent systems in India and high out-of-quota tariffs in China are expected to remain following the Doha Round under Special and Differential Treatment for developing countries. Relatively low tariff barriers mean that any reduction will have relatively small effects on the location and scale of production or on trade flows.

Several studies point in this direction. A study by Promar International for the United Soybean Board concluded that a 50% reduction in remaining trade barriers would result in only a 0.4% increase in world oilseed production and use. A study by Agriculture and Agri-Food Canada came to virtually the same conclusions: world oilseed production would increase 0.2% and prices would rise by 2.1%. Both studies found the greatest effects to be on the oil side, with increases in vegetable oil trade and a strengthening of oil prices relative to meal.

The United States Department of Agriculture (USDA) study examined the production and trade implications of full trade liberalisation. For the oilseed sector, it concluded that complete elimination of all agricultural supports and protections worldwide would increase total world production of oil and oilseeds 0.7%. Output would decline 5.3% in developed countries and rise 4.8% in developing countries. World trade volume would go up 11.4% A separate USDA study that year, using a different model, looked specifically at the oilseed sector and the implications of eliminating all import duties and export taxes—the so-called "zero-for-zero" option. It found that oilseed output would rise by 0.4—1.8% in major exporting countries, and that Malaysian palm oil output would rise 4.1%. Again, trade in oils would increase more than trade in oilseeds or oilseed meals.

More recently, the Food and Agriculture Organisation (FAO) analysed two different levels of ambition in the Doha Round—one roughly similar to the Uruguay Round cuts, and one with bigger Swiss formula tariff cuts, a 50% expansion in tariff rate quotas (TRQs) and an 80% cut in domestic support. Impacts on oilseeds under the first scenario are limited: trade expansion of 1% for oilseeds and 6% for oils, with price increases of 4.2% and 3.2%, respectively. The second scenario resulted in a 3% gain in oilseed trade, an 18% gain in oil trade, and price increases of 10% and 8%, respectively. Because the major markets for oilseed meal are highly distorted, the study could not predict the potential for liberalisation in that sector.

These aggregate changes are similar in magnitude to changes that are occurring already on an annual basis due to rising global demand. For example, world consumption of vegetable oils has grown at an average annual rate of 4.4% since 2000/01, and 4.6% over the past decade. Consumption of oilseed meals has grown at a 3.4% rate since 2000/01, and 4.8% over the decade. In addition, the oilseeds sector is changing rapidly even without changes brought about by multilateral trade agreements. The geography of oilseed production has been shifting in the last decade to South America and Asia. Demand has been shifting as well, changing focus from Europe to Asia. Thus distinguishing the output effects of trade liberalisation from those related to the ongoing rapid growth in demand and shifts in supply sources is a challenging task.

Soybeans

Soybeans and soybean oil are consumed worldwide, but six countries account for about 95% of world soybean production: United States, Brazil, Argentina, China, India, and Paraguay. Production in India and China is almost entirely for domestic consumption, and the remaining four countries account for essentially all exports of soybeans and soybean oil, which accounts for 30% of global production. As a result of cost competitiveness and the tremendous potential for growth in output, the production and processing of soy for export has been rapidly shifting from North to South America.

Low land costs and the devaluations of the Argentine and Brazilian currencies have contributed to the large production cost advantage of South American producers compared to North American soy producers. Since 1996 the US government has increasingly supported its soybean sector through subsidies meant to serve as a financial safety net for producers. These subsidies, which include export assistance, have affected global prices of soy and are believed by some to have contributed to the chronic depression of vegetable oil prices in recent years. And, while tariffs have been relatively low in the soy sector, non-tariff barriers have become more controversial with regards to state trading enterprises and labelling requirements for genetically modified products.

The controversy over the use and acceptability of biotechnology has had major ramifications for trade in oilseeds and oil-containing products. Mandatory traceability and labelling regulations by the EU have accounted in part for some of Brazil's increased exports to the EU and for the EU's increase in palm oil imports. Europe and Asia are the world's largest importers of soybean, either in their raw form or as processed oil and meal products. Because most soybeans are crushed to yield about 80% meal and 20% oil, the soy export market is driven primarily by the demand for meal and less by that for oil.

The meal is very high in protein, and a valuable additive for animal feeds. Demand for soybeans has increased by 64% in the past ten years, but change has been most dramatic in China and India where meat and vegetable oil consumption are expanding rapidly. These two countries now account for about one-third of world soybean oil use and the United States for a quarter.

The major oilseeds are typically grown by farmers who also grow grains and other field crops. Soybeans are a row crop, with mechanised planting in the US in rows about a half-meter wide and seeds about 5 centimeters apart. Planting methods have included closer cropping methods in Brazil and other countries where climatic conditions differ. In most areas, cultivation, fertilizer and pesticide application, and harvesting are mechanised.

While soybeans in temperate regions are typically rotated with corn or other crops on an annual basis, they can be rotated biannually in tropical regions where two crops can be produced in one year. Environmental concerns associated with soy production are generally the same as for other mono-cultured oilseed crops. They include chemical application, soil and water contamination, monocrop expansion into environmentally sensitive areas, leading to loss of wildlife habitat and soil degradation, and genetic modifications that could promote superweeds or other environmental problems.

The use of fertilizers and crop protection chemicals in soy crops is low compared to other field crops, and differs with respect to the kind of cultivation used and region of growth. No-till methods, also called conservation tillage or direct planting, are generally environmentally preferable because they leave more organic matter on the soil surface, reduce erosion and runoff, reduce machinery use and lower the costs of production.

Biotechnology has encouraged no-till farming and soybeans have been one of its prime beneficiaries. Herbicide resistant soybeans can be planted directly into the ground without tillage of the soil and then sprayed with herbicides, leaving the edible bean plants intact. This reduced input costs for farmers, who do not need to apply the volume and variety of herbicides used with conventionally planted non-resistant varieties. It has also reduced soil erosion and improved soil and water quality. No-till planting methods can also be used with conventional (non-GM) crops in some regions.

The processing of soybeans can also have impacts on the environment—those associated with processing in general and with chemical refineries in particular. These include effluent, energy use, and air pollution, and are associated both with the refining process and transportation. Crushing has relatively small impacts in developed countries where wastewater is typically treated in municipal water treatment systems, but wastewater loadings may be an environmental issue in developing countries. Solvent extraction of soybean oil from the flakes can result in atmospheric emissions of hexane.

Social problems posed by soy culture and processing in developing countries are generally similar to those encountered in many other agricultural development enterprises. They can include exploitation of workers and disruption of indigenous communities, exposure to health hazards caused by agricultural chemicals and machinery, and unsafe working conditions. The scale of soybean production can also exacerbate the rural-urban transition as displaced workers cease production and head to cities for employment.

The industry structure in the soy value chain differs from country to country, but in most countries large farms, or cooperatives or small farms unified by efficient transport schemes, can best produce soy to the economies of scale that now characterise the industry. The soybean trade in, transport to, and ownership of crushing facilities, is concentrated in a few large multinationals. Cargill, Bunge, ADM, Louis Dreyfus and Glencore account for more than 75% of global crushing and are highly competitive in all major markets.

The multinational compound feed industry, which further processes soy meal for animal feed, is regionally based while refined soy oil for food use is divided between global players and family-owned refining and bottling companies. In Latin America, soy producers are often financed by crushers who can also provide social benefits that might not otherwise be locally available. The industry also has access to credit from large commercial banks and has used highly developed risk management practices. Rapid price increases in soybeans have yielded new sources of credit availability in many countries.

Palm Oil Industry

Within the oilseed sector, consumption of palm oil has been growing faster than that of soy or rapeseed, primarily because of its lower cost of production. No other oil comes close in terms of growth rate. In addition to low production cost, palm oil is also used as a primary oil in geographical areas where population growth and economic development are rapid (India and China), and it has partially supplanted soy oil in Europe due to the region's opposition to genetic modification of soy.

Palm oil is semi-solid naturally and does not require hydrogenation. It can be separated into liquid and solid fractions, referred to respectively as olein and stearin, and either of these can be traded in their crude or refined forms, as can the oil from palm kernels, the seeds within the palm fruit. Malaysia and Indonesia account for about 85% of all palm oil production, and most of this is destined for export. Indonesia is expected to overtake Malaysia as the world's largest producer of palm oil between 2012 and 2015, yet its concerns for and commitments to selfsufficiency in domestic oil consumption (mainly for cooking) will ensure a long-term lag in export volume relative to Malaysia.

Nigeria is also a large producer with 30% of the world's total plantation area, but normally its entire output is consumed domestically. Expansion of palm oil production has begun in some parts of Africa and Latin America but has not yet reached levels competitive with Southeast Asia. Most palm oil production is exported and therefore producers are highly exposed to world market prices. Because there are a number of substitutes for palm oil, the market is highly competitive.

Markets for byproducts of other oilseeds also influence this market. The main market driver for soybean demand, for example, is soy meal, used primarily for animal feed. Thus, price changes in the feed industry send strong signals to the palm oil industry due to a high cross price elasticity. The energy market is also taking great interest in the potential for alternative fuels, namely biodiesel and ethanol, implying that oil palm producers may be more directly affected by energy prices, specifically petroleum prices, in coming years.

The majority of palm oil trade is within Asia. Exports from Malaysia and Indonesia are particularly large to India (18% of world imports), China (12%), and Pakistan (17%). The EU also draws heavily on palm oil imports and now accounts for 17% of world imports. Malaysia and Indonesia are expected to gain from trade liberalisation initiatives, given their dominance as exporters on the world market. Reductions in oil import tariffs in importing countries would make palm oil cheaper to consumers in those countries and increase demand in the international market.

India, the world's largest importer of vegetable oils and the single largest buyer of palm oils produced by Malaysia and Indonesia, imposes high bound tariffs of 300% on most edible oils. The major exception, soybean oil, is subject to bound tariffs of 45%, which has boosted soybean oil imports over palm oil. Palm oil has had to be sold at a discount in order to remain competitive in a highly price sensitive environment and Malaysia has been calling for a more coherent policy in India.

Exporters would benefit greatly if India reduced its palm oil tariff ceiling. In the European Union, palm oil consumption has increased by 140% since 1994/95 to 13% of global usage. Yet the EU has maintained a disparity between tariffs on crude palm oil and those on refined and processed products coming from Malaysia and Indonesia. The EU also continues to offer preferential access to several palm oil producing countries, including Papua New Guinea, Ivory Coast, Nigeria, Ecuador, and Colombia.

Oil palms are large trees, typically planted in grids on plantations that range from 4 to 729 square kilometers. Creating plantations requires removal of standing vegetation by cutting, mechanical clearing or burning. Grids are often planted with little regard to topography, which sometimes results in planting on steep slopes. Land is prepared for planting by mechanically weeding or treating with herbicides. The trees start to flower in 2-3 years and can produce for as many as 50 years, but new propagation methods are decreasing the time from seedling to maturity and enhancing production, while reducing net life-span. Peak production of most areas planted now is from 8 to 30-year-old trees.

Fertiliser is important to maintain production, and input costs can range from 40 to 60% of total maintenance costs. However, new techniques, such as underplanting of nitrogenous crops, can reduce the need for fertilizer. Until the canopy closes and shuts off light, cash crops can also be planted. Establishing new palm oil plantations requires significant capital investment, first for the processing facility, second for enough land planted to oil palms to service it, and third for the money tied up in the plantations until they start to produce.

Mills typically require at least 10,000 hectares of productive land to be viable. This often creates a symbiotic relationship between large and small producers, in which smaller producers surround large-scale producers that typically invest in crushing mills in order to meet their processing needs. This has been changing with the emergence of smaller and more mobile processing mechanisms. In Malaysia and Indonesia, oil palm plantations are typically established on converted forest land or converted rubber plantations. These occupy a large part of peninsular Malaysia and are increasing in Indonesia due to Malaysian and Chinese foreign investment.

The environmental consequences of producing palm oil have been considered as the loss of environmentally sensitive forest habitat. The growth of the palm oil sector continues to have significant impacts on forest resources in high value tropical forests. Clearing the land for palm oil plantations often involves illegal burning, and policies in some areas have provided perverse incentives for deforestation, rather than replanting abandoned agricultural land. Recently, producers and some important end users have come together to discuss a set of best practices and quality assurance guidelines that could govern both production and purchasing decisions.

In an initial "Roundtable on Sustainable Palm Oil", some specific environmental concerns were identified. These include:

— Forest conversion

— Fire from land clearing, causing extensive air pollution and haze

— Planting considerations (on steep slopes, which are prone to erosion, and on peat soils, which are difficult to cultivate sustainably)
— Biodiversity on forest plantations, which is lower than that of native habitats
— Agronomy and mill management, and
— Lack of enforcement of environmental laws, specifically those outlawing burning of forests to clear land.

Rapeseed Oil

In general, rapeseed oil is used for industrial purposes because of its toxicity. For human consumption it must be altered to a higher-value oil and protein crop known as canola. Canola meal can be fed to livestock, while use of rapeseed meal as feed is limited to cattle because of its glucosinolate content. Very little rapeseed or canola oil enters world trade, so consumption tends to be concentrated in the countries where the crop is grown. In Europe, the biodiesel market has become one of the main drivers of the rapeseed market, consuming about one third of EU-25 rapeseed oil production.

Approximately 80% of biodiesel production comes from rapeseed oil. Rapeseed and canola are prolific cross-pollinators, and much care is taken to strictly separate food and industrial varieties. Work in Europe on the environmental effects of rapeseed production has focused on the effect of increased production in sequestering carbon emissions. Like other row crops, the purely environmental effects of rapeseed cultivation are generally acknowledged to include loss of biodiversity due to cultivation, soil degradation and effects of fertilizer and crop protection chemical use. There are several ways to minimise these effects including:

— Minimizing the amount of set-aside land used for rapeseed (biofuels) to maintain biodiversity and encourage natural land regeneration,

— Minimising the use of artificial fertilizers for biofuels to reduce nitrous oxide produced through excess fertilizer decomposition,

— Using low levels of herbicides and pesticides to keep weeds and pests within acceptable limits, rather than eradicating them completely,

— Interspersing rapeseed cultivated areas with other crops and sowing different parts of the crop for harvesting in different seasons, and

— Retaining uncultivated buffers around rapeseed areas, such as at field edges or around watercourses to maintain plant biodiversity and promote natural regeneration of the land after cultivation.

Changes to the Sector

Over the last decade world demand for protein meal and vegetable oil grew at annual rates of 3.7% and 4.4%, respectively, according to USDA figures. Oilseed production grows at about the rate of meal demand but obviously varies from year to year due to weather. Looking ahead, a conservative forecast based on past trends might cut these growth rates by a fifth to 3.0% and 3.6% respectively on the theory that the most dramatic impact of China's economic changes is past. That still leaves 6-7 million tons of additional protein meal per year, and 4 million tons of additional vegetable oil per year.

USDA's February 2005 Long-term Baseline projections do not include numbers for total oilseed sector demand growth, but they do include a trade forecast for soybeans and the soybean-equivalent of soybean meal that implies continued strong growth in world demand. To produce an extra 6-7 million tons of protein meal each year requires about 10 million tons of oilseeds, and soybeans have typically been about 60% of the total. For oil production and use to grow faster than meal demand, production of palm oil or high oil content seeds like rapeseed and sunflowerseed needs to grow faster than soybean production. Historically it has been palm oil that has shown the fast growth.

Some growth in oilseed output has come from the upward trend in yields, which have grown at 1.2-1.8% per year depending on the oilseed. But growth in demand has been so much more rapid than for other major field crops that planted area has also had to increase. This new acreage is partly due to shifts from other crops and partly to bringing new land into production. Oil World estimates that in the 20 years ending in 2002/03, harvested oilseed area rose by an average of 2.75 million hectares per year. In the following two seasons it rose an average of 10 million hectares per year, far above the historic average rate. For 2005/06, world oilseed area harvested is estimated to be essentially unchanged from a year earlier.

Production of Biodiesel

Rising demands for renewable energy must be factored into any discussion of the future of the oilseed sector. During the last quarter of the 20century, there have been two separate renewable energy initiatives for biofuels. Fuel ethanol has gained a foothold in both the United States and Brazil either blended with gasoline or used as a straight automotive fuel. In the United States the raw material for ethanol production is primarily corn while in Brazil it is sugarcane, and both countries have developed significant production capacity for their respective crops through government support.

In Europe, where diesel powered-cars are more popular, biodiesel has gained wider acceptance. Europe's demand for biodiesel has been a driving factor in its support for rapeseed production. In practical terms, the three oils of interest for biodiesel are soy, palm and rape, either because they are cheap or because they are produced in the countries that can afford to subsidize these more expensive fuels. In recent years a number of countries outside of Europe have been supporting research and development programmes to find new end-uses for specific oilseeds and derived products. The regular provision of subsidies and/or tax breaks to refiners has been necessary to guarantee the economic viability of bio-diesel production from oil crops, although the fuel price increases that have occurred in 2005 may obviate that need.

The future demand for biodiesel has the potential for reshaping the dynamics of the oilseed industry, with a greater emphasis on vegetable oil as an industry driver. Because of commitments to reduce greenhouse gas emissions under the Kyoto Protocol, and a desire to reduce dependence on imported fossil fuels, several countries have adopted ambitious renewable fuel strategies. Subsidies and tax credits applied to alternative fuels will likely fall into the Green Box category of policy tools under the WTO, in which case they would not be subject to substantial reform. As the price of petroleum rises, the EU in particular will begin to look beyond their limited stocks of oilseeds and other field crops to satisfy growing demand.

The European Union's Biofuels Directive set targets for a minimum share of transport fuels for biofuels: rising from 2% in 2005 to 5.75% by 2010. A subsequent directive in 2004 encouraged the development of a biodiesel industry by allowing member states to "de-tax" (or reduce taxes on) fuels from renewable sources. In early 2004, Rabobank forecast biodiesel consumption at around 4-6 million tons by 2010 in the EU-25. Subsequent world oil price developments suggest the upper end of that range is more likely.

In addition to being used to produce biodiesel, vegetable oils are also being blended directly with petrodiesel, used in home heating systems, and burned to produce "green electricity" in Europe. As a result of its aggressive renewable fuels policy, the EU had total biodiesel production capacity of almost 2.4 million tons in 2004 according to the European Biodiesel Board (UFOP). Actual production in that year is estimated at 1.4 million tons, with Germany accounting for about 60% of the total. Biodiesel production in the EU will expand to a projected 6 million tons in 2010, with rapeseed oil the primary raw material, but with substantial quantities of imported palm and soybean oil being used as well.

The rapid expansion of the biodiesel sector in Europe means that rapeseed oil will be supplemented by soybean oil from the Western Hemisphere and palm oil from Malaysia and Indonesia, both of which can soon enter the equation. Soybean oil is in surplus in Europe because of consumer aversion to genetic modification. Palm oil's position as the world's cheapest oil puts it in a good position to capitalize on Europe's growing appetite for alternative fuels. Recognizing this, the Malaysian government has proposed a policy to promote the conversion of palm oil into biodiesel.

Malaysia itself and Asia as a whole are also expected to use palm oil as energy policies and crude oil prices drive industries to favour the use of diesel blends. There is also likely to be increased production in of rapeseed Europe. The renewable fuels policy has already increased rapeseed production in the EU even though support has been reduced under Common Agricultural Policy (CAP) reforms. Since current production for biodiesel constitutes approximately 2.7 million tons of the total oilseed production of 13.1 million tons, future increases could consume a substantial percentage of total oilseed production.

For instance, French production of rapeseed is projected to more than double by 2007. This is ultimately projected to reduce demand for soybean meal, as rapeseed meal is substitutable in animal feed at a rate of 1.5 pounds to 1 pound of soybean meal. Under the current CAP, the arable raw materials needed for biodiesel production may be grown on set-aside land, which would otherwise be taken out of production. The United States and Brazil are likely to be the next two largest biodiesel markets after Europe.

In the United States, the Energy Policy Act of 2005 combines a renewed tax credit with a renewable fuels standard (RFS) to require a percentage of biodiesel in fuel. The bill is expected to expand the use of biodiesel to 7.5 billion gallons. Biodiesel will be a more important factor for Brazil than for the United States because in addition to its domestic use, Brazil will be the main beneficiary of EU import demand for soy for biodiesel, in the form of either beans or oil. Canola oil from Canada may also contribute to this feedstock, since Canada's domestic demand for biodiesel is projected to be low.

Canada's Climate Change Action plan targets 500 million litres/year of capacity by 2010. Malaysia and Indonesia are together expected to provide up to 20% of Europe's supply. Estimated world biodiesel production in 2010 is likely to be composed of 4.5 million tons of rapeseed oil, 1.9 million tons of soybean oil, 1.2 million tons of palm oil, and 150,000 tons of other oils. The expansion from 2004 levels will require 9.3 million tons of rapeseed, 9.2 million tons of soybeans and 5.75 million tons of palm fruit bunches.

These are rough estimates based on secondary sources and are intended only to capture the order of magnitude of the change underway. Much will depend on how energy prices and the world economy evolve over the next few years, and on interactions within the sector, such as rapeseed meal displacing soybean meal. The key conclusion, however, is that over the next six years, the growth in oil used for fuel will be a significant increment over the historic growth pattern.

The growing demand for human consumption and industrial purposes will require an additional 4 million tons of oil production per year, or an extra 24 million tons of oil production over and above the 2004 base consumption of 105 million tons by 2010. Add to that the projected annual demand of 6 million tons of oil for biodiesel and the total requirement for 2010 amounts to about 135 million tons of oil. Meeting this demand will require a 29% increase in production.

Taking the high end of the range, the Doha Round might expand world production by 1.5% over a number of years. By 2009/10 world oilseed production will be about 410 million tons in the absence of biodiesel expansion and trade liberalisation. As noted above, biodiesel will require an additional 18.5 million tons of rapeseed and soybeans, or about 1.5 years of normal growth in production, plus 1 million tons of palm oil. Trade liberalisation will have a considerably smaller effect—about 6 million tons of oilseeds plus something like 1.2 million tons of palm oil.

The two together could push world oilseed production that year to 435 million tons and palm oil production to 42 million tons. Cross-commodity effects between the soy and palm sectors will occur but it is difficult to separate those stemming from trade liberalisation, biodiesel demand, and underlying trends in the sector. The high protein content soy sector and high oil content palm sector are already closely linked through the overall dynamics of food demand. Increased oil demand due to biodiesel and trade liberalisation would tend to tilt the balance toward palm, and that trend could be strengthened by soy's liabilities: the negative perceptions of trans fatty acids and of biotechnology.

Another background factor adversely affecting soy is the competition for soybean meal in the United States from mid-protein by-products of fuel ethanol production, such as distillers dried grains and corn gluten feed. There are now about 10 million tons of these byproducts being produced annually that will expand considerably over the next few years. However, trade liberalisation will enhance the demand for livestock products, and the derived demand for soybean meal and other feedstuffs.

Biotechnology

While some of the biotechnology debate is academic and hypothetical (particularly relating to the environmental effects of varieties yet to be developed and used), some of the positive environmental and economic effects resulting from using the new biotech-derived seeds are evident to producers, who have embraced the technology in most areas where it has been permitted. Successive studies have noted reduced use of pesticides and reduced insect and pest damage, and consequent reduced input costs for agrichemicals.

No-till methods account for some of the general improvements in soil condition attributed to biotech crops, but biotech varieties have also made no-till methods profitable in some areas where conventional varieties did not allow for this method of cultivation. Biotechnology has also permitted increases in yields that have been fairly well-documented. Use of biotechnology derived crops increased from 2003 to 2004 by 20%, amounting to 13.3 million hectares. However, for countries whose production is export-oriented, biotech production is only economically beneficial if it can be exported. The EU's regulations, particularly as they have been replicated elsewhere, may make biotech production for export less feasible for many countries.

REFERENCES

Amiruddin, Mohd, N., "Palm oil products exports, prices and export duties: Malaysia and Indonesia compared," *Oil Palm Industry Economic Journal,* Vol. 3 No. 2: 21-31, 2003.

Basiron, Y., N. Balu, and D. Changrmohan., "Palm oil: the driving force of world oils and fats economy," *Oil Palm Industry Economic Journal,* Vol. 4 No. 1: 1-10, 2004.

Casson, A., "Oil palm, soybeans and critical habitat loss," *A review prepared for the WWF Forest Conversion Initiative,* August, 2003.

Chand, Ramesh, 2004, "WTO and Oilseeds Sector: Challenges of Trade Liberalization", *Economic and Political.Weekly,* February 7, 533-537.

Dohlman, E., S. Persaud, and R. Landes, 2003, "India's Edible Oil Sector: Imports Fill Rising Demand", OCS-0903-01, *Electronic Outlook Report from the Economic Research Service,* USDA.

Gulati, A., A. Sharma, and DS Kohli, 1996. "Self-Sufficiency and Allocative Efficiency: Case of Edible Oils", *Economic and Political Weekly,* 31(13): A14-A24.

Niesten, E.T, Rice, R.E., Ratay, S.M. and Paratore, K (eds.)., "Commodities and Conservation: The Need for Greater Habitat Protection in the Tropics," Washington D.C., Conservation International, 2005.

Young, C.E., and P.C. Westcott, 2000, "How Decoupled is U.S. Agricultural Support for Major Crops?", *American Journal of Agricultural Economics,* 82 pp. 762-767. 34

10

Industrial Challenges of Oilseed Crops

Over the last few decades, the world has become dependant on fossil fuels as the primary source of hydrocarbons for energy and industry. These resources are finite and non renewable and their exploitation releases massive amounts of carbon dioxide into the atmosphere. The identification of sustainable replacements has been an ongoing area of research for many years.

Plant triacylglycerol (TAG) oils have shown considerable potential to replace petrochemicals in a wide variety of applications ranging from fuel and lubricants to the manufacture of polymers, surfactants, and surface coatings. Their utilisation by these industries has however been limited by a number of factors, the most important being the inability of vegetable oil production to compete with the low price of mineral oil. The vegetable oil industry has therefore focused on producing oil for food and food processing uses.

With the recent rapid increase in crude oil prices and the focus of political attention on renewable energy and bio-sustainability, the market for vegetable oils is beginning to undergo some fundamental changes. Currently this is lead by the demand for vegetable oil for the manufacture of biodiesel, as illustrated by the more than 3 fold increase in biodiesel production in the US from 25 million gallons in 2004 to 78 million gallons in 2005. It is important to realise that although biodiesel production has clear environmental benefits it is not a viable replacement for fossil fuels in transportation and energy usage. The reason for this is that it is simply not possible to produce enough oil to substitute for more than a fraction of our current petroleum consumption. An area where vegetable oils show long term potential, either as TAG or as a source of fatty acids, is in higher value applications such as lubricant manufacture and as industrial raw materials. There are, however, considerable challenges that must be overcome for plant oils to achieve their full potential for many industrial processes.

SCIENTIFIC CHALLENGES

The chemistry of plant oils offers both advantages and disadvantages to industrial processes. The fatty acid composition of the oil determines its properties and its utility. Most plants only produce a small number of fatty acids, either 16 or 18 carbons in length with zero to three double bonds at specific positions. These are often referred to as the "common" fatty acids and they are the predominant fatty acids of commercial vegetable oils, a market dominated by only 4 species; soybean (*Glycine max* L., Fabaceae), oil palm (*Elaeis guineensis* Jacq., Arecaceae), rapeseed (*Brassica napus* L., Brassicaceae), and sunflower (*Helianthus annuus* L., Asteraceae).

Within the plant kingdom, however, a diverse array of "unusual" fatty acids are synthesized, many of which would be potentially valuable industrial raw materials if available in sufficient quantity. These fatty acids differ from the common fatty acids in aspects such as chain length, degree of unsaturation and double bond position, or by the presence of functionalities such as hydroxyl, epoxyl, and acetylenic groups, and even covalently bound halogens. Although these fatty acids can be present at high levels in seed oils they are usually only found in a small number of plant species, the majority of which do not have agronomic characteristics suitable for their development as commercial oil crops.

Since the development of molecular biology techniques, considerable effort has been directed at identifying the genes encoding enzymes responsible for synthesis of unusual fatty acids. The goal was to transfer these to existing oilseed crop plants to produce novel oil for industrial uses. In most instances results have been disappointing, with only low levels of unusual fatty acids being produced in the engineered crops. The target fatty acids were also found both in the TAG and the membrane lipids of the transformants whereas in the native species they were only found at significant levels in TAG.

Incorporation of certain unusual fatty acids into membrane lipids is thought to be detrimental to membrane function and the native plants have developed mechanisms for their exclusion. The challenge is now to understand and overcome the factors that are limiting the accumulation of these fatty acids in transformed plants. Part of the solution will be to gain a better understanding of the processes of TAG assembly, particularly in plants that make unusual fatty acids, but also in existing oilseed species.

It is likely that multiple factors are responsible for high level production of unusual fatty acids and there is no single gene that will allow this in a transgenic oilseed. Furthermore, when higher levels are achieved, it may also be necessary to modify the breakdown of unusual fatty acids during seed germination so that the engineered crops are able to efficiently utilise the novel fatty acids that they are synthesising.

Current strategies to increase unusual fatty acid content are now focusing on "pathway engineering," the cloning of biosynthetic enzymes and enzymes involved in TAG assembly from species of interest and their cotransformation into the target oilseed. There is increasing evidence that enzymes of TAG assembly from plants that make unusual fatty acids have high specificity for these products whereas homologous enzymes in the target crops show a preference for the common fatty acids.

An alternative to the transfer of fatty acid biosynthetic genes to existing crops is to apply molecular biology techniques to aid in the development of new crops, or to improve the characteristics of existing minor oil crops that produce the target fatty acids. Techniques such as Tilling or RNA interference can be applied to silence specific genes in a plant and the development of molecular markers can greatly speed up breeding programs.

The major obstacle to this approach is that agronomic traits are generally complex and often not well understood. Application of these techniques to improve specific characteristics such as fatty acid composition, or to reduce the content of undesirable seed components, such as ricin in castor bean (*Ricinus communis* L., Euphorbiaceae) may however have potential.

Increasing the portfolio of fatty acids available to industry may be more successful in the short term for fatty acids that have value even if they are not the major component of the oil. The presence of relatively low levels (10% to 15%) of ricinoleic acid (12-hydroxy oleic acid) in a vegetable oil high in oleic acid, for example, has been shown to greatly improve the characteristic of the oil in lubricant applications. It has already been shown that these levels can be achieved by transformation of oilseeds with the oleate-12 hydroxylase gene from castor bean. Programs are currently underway to develop a commercial crop that produces a seed oil containing 15% ricinoleic acid (Linnaeus Plant Sciences Inc., unpubl.).

Vegetable oils are generally heterogeneous in composition with each TAG molecule containing a number of different fatty acids. As an industrial feedstock this mixture is not ideal. In processes that involve the hydrolysis of the oil, fractionation steps will be required to isolate the fatty acids of interest and this adds to the processing costs of the material.

Although there may be value in each of the fatty acids, the most suitable oil would be homogeneous containing a very high percentage of a single fatty acid. Considerable progress has been made in recent years in increasing the homogeneity of fatty acids that naturally exist in the major crop species.

A good example is the increase in oleic acid (18:1) content achieved through conventional and mutational breeding of canola (*Brassica napus*), soybean, sunflower, safflower (*Carthamus tinctorius* L., Asteraceae), and peanut (*Arachis hypogaea* L., Fabaceae) where cultivars containing this fatty acid in the range of 80% to 90% have been generated. Similar levels have been achieved in canola, soybean, and cotton by genetic engineering. Although developed for food uses, the availability of a relatively pure source of oleic acid at commodity prices has stimulated interest in the industrial uses of these oils.

Breeding strategies have also been applied to increase the content of linolenic acid (18:3) from around 45% to over 65% in flax (*Linum usitatissimum* L., Linaceae) and erucic acid (22:1) to similar levels in rapeseed, making these oils more suitable for their role as industrial feedstocks. Biological constraints however appear to exist which may limit the accumulation even of naturally occurring fatty acids in some seed oils. Erucic acid for example in rapeseed is not efficiently esterified to the center (*sn*-2) position of the TAG molecule. The theoretical maximum limit of erucic acid in this species would therefore be only 67% of total seed fatty acids.

Efforts to overcome this barrier by overexpressing the condensing enzyme responsible for erucic acid synthesis in conjunction with an acyltransferase enzyme capable of catalysing the incorporation of erucic acid into the *sn*-2 position, have been successful in altering the distribution of the fatty acid in TAG, but have not significantly increased the total amount in the seed. This fatty acid, and 18:3 in flax, are the end products of a pathway of fatty acid modification in the developing seed. High level accumulation would therefore be expected to require efficient conversion of substrate in addition to efficient incorporation of the product into TAG.

TECHNICAL CHALLENGES

In addition to the scientific challenges associated with developing vegetable oils with new fatty acid profiles there are a number of technical issues that must be considered. The choice of fatty acids will be determined by their utility and also by the size of the market for the oil. This is complicated by the fact that the markets may not yet exist as the product is essentially a new chemical feedstock.

Rather than trying to introduce too many novel fatty acids it would be more logical to restrict the number to a few key substrates that can be further converted to the desired feedstock. In this respect monounsaturated fatty acids may be of particular value as the single area of reactivity will allow more control over chemical modification. The presence of the double bond also allows the application of olefin metathesis chemistry to produce dicarboxylic acids and alkenes.

Identity preservation will be a key issue and seeds destined for industrial uses will require segregation from those grown for food use. Mechanisms must be available to control all stages of production and also to ensure that traits are not accidentally transferred to other varieties or species. Possible approaches to simplify this include the engineering of existing major oilseed crops, but with geographic separation during production, or the establishment of platform crops that are not widely grown, but which have the potential to be developed into industrial oilseeds. For temperate zones, candidates include members of the Brassicaeae such as *Brassica juncea* L., *Brassica carinata* A. Braun, *Camelina sativa* L., and crambe (*Crambe abyssinica* Hochst. Ex. R.E. Fries). The advantage of a species such as *B. carinata* is that it can grow well on marginal land that is less suitable for the production of food crops such as *B. napus*.

For oilseed engineering to be successful it is important to consider all aspects of the seed, not just the oil. A co-product in vegetable oil production is the meal left over after oil extraction. Uses will have to be found for the meal and this may be influenced by the nature of any residual fatty acids and the presence of anti-nutritionals such as glucosinolates. The ability of seeds to germinate and establish should not be affected by the novel fatty acids and any impact on oil content would need to be offset by an increase in value.

For some countries it may also be important to breed in agronomic characteristics such as herbicide resistance. The new crops should provide valuable raw materials for industry but also bring value to the producers, and this means the farmers Industrial Oilseeds who are growing the crops. Perhaps most importantly, large scale development of industrial crops must not compromise the ability to grow food.

Vegetable oils have a bright future as a source of renewable raw materials for industry. Considerable effort is still required, however, to complete our understanding of how plants synthesize oils of particular fatty acid compositions and how these processes can be modified to optimise vegetable oils for specific purposes.

MATRIX APPROACH TO PRODUCTION OF OILCROPS

The purpose of the matrix approach is to derive the

environmental sustainability implications of changes in scale and location of agricultural production that will accompany agricultural trade reform. It does so by illustrating in chart form the environmental repercussions of changes in agricultural production associated with the terms of international trade. In one dimension the matrix lists the relevant causal policy measures, and in the other the environmental indicators that show effect. This matrix approach originates from a paper; "Trade, Agriculture and the Environment" co-authored by John A. Dixon and M. Ann Tutwiler, and makes it possible to indicate the kinds of environmental problems that are likely to arise from explicitly distorting policy measures and the underlying (implicit) policy measures that may contribute to them.Our matrix, however, differs in some respects from the one used by Dixon and Tutwiler. In its treatment of implicit and explicit distortions the matrix includes,

— explicit domestic policy measures relating to biodiesel. We included this because of the great influence that biodiesel-related policies will have on growth of planting areas in the countries studied, and consequently on environmental indicators.
— a focus on the three major components of the Doha negotiations (market access, domestic support and export subsidies) because focusing on the "boxes" at this time would lack clarity in light of recent WTO Appellate Body rulings and the state of the negotiations.
— direct input subsidies in domestic support.
— the enforcement of environmental regulations/rule of law, which is treated as an implicit distortion rather than an explicit one, and considers undefined property rights. The treatment of environmental effects, combines soil mining and erosion into a "soil erosion/ quality" category,

- combines groundwater and surface water contamination into a "water quality" category,
- eliminates a category for "loss of other productive use" because good information sources are lacking, and
- creates a "social impacts" category to take into account not only deleterious human health impacts but also urban migration of rural populations, disruption of indigenous communities, general working conditions and job creation, A brief description of each category in the matrix follows, together with the question that each poses for oilseed production vis-à-vis the environment of each country.

POLICY MEASURES, EXPLICIT AND IMPLICIT DISTORTIONS

Biodiesel

For reasons that should be obvious at this point, biodiesel has been included as an explicit policy distortion because its development is being driven primarily by government policies that either mandate or subsidize its use. Since the scale of biodiesel's impact is likely to be considerably larger than that of trade liberalization, it is important to keep this in mind as a background factor as one looks at the relative effects of the different components of trade liberalization or implicit distortions.

Market Access

Both tariffs and quotas are being negotiated in the Doha Round. Because import tariffs and quotas are market distorting, high tariffs (and low quotas) impeding market access are potentially disruptive to environmental improvements unless they are associated with measures designed to compensate for environmental externalities not included in the prices of a product. In the agriculture context, high import duties are also associated with either protected or value-added products. Protecting production through high tariffs may promote environmentally disruptive production at the expense of market signals. Applying a higher duty for processed products than for unprocessed ones can encourage production of basic raw commodities at the expense of processed ones, which may save on the environmental costs associated with a heavily polluting processing industry but may also cause environmental degradation if it encourages the use of non-sustainable farming practices to maximize volume.

The Doha Round is expected to cut high tariffs more than low ones, thus reducing or eliminating tariff escalation, but developing countries will receive special and differential treatment that will most likely result in longer phase-out periods for tariffs and end rates of less than zero. "Sensitive" products may also remain protected. Rates in the oilseed sector are already rather low in comparison to other agricultural commodities, but there is a great gap in some countries between bound and applied rates.

Non-tariff Barriers

Non-tariff barriers are relevant to production of all agricultural commodities. In the oilseeds sector, such barriers include traceability and labeling for commodities derived from modern biotechnology, sanitary and phytosanitory (SPS) requirements relating to plant pests and diseases such as soybean rust (US) and government-mandated requirements such as import certificates (China). The SPS and technical barriers to trade (TBT) texts are not expected to be negotiated in the Doha Round, but there may be increased levels of technical assistance given to developing countries in this area.

Domestic Support Measures

Domestic support measures linked to production (Amber Box) have been criticized as distorting price competition, limiting imports and creating oversupply, making exported commodities cheaper in the importing country than domestically produced ones. Supports not linked to production (Green Box) and production-limiting compensation (Blue Box) have also been criticized in the US and EU because of their continued high levels and implicit impact on trade.

The United States and the European Union have maintained high levels of support for their farming sectors, of which the oilseed sector has been a primary beneficiary. The Doha Round is expected to substantially reduce Amber Box payments.

Blue Box payments would likely be capped at 5% of the value of production. This would result in a relatively significant change for developed country oilseed producers and might shift production to other crops for which production-related payments are not likewise disciplined, in the absence of market signals to the contrary.

Continued strong demand fueled by biodiesel and high prices, for example, could keep oilseeds competitive. Green Box compensation may also offset oilseed disciplines in other categories through support for biodiesel and may even increase production of oilseeds. Blue Box compensation can be harmful to the environment if it keeps marginal or environmentally sensitive land in production.

Export Subsidies/Taxes

Export subsidies can affect the environment through terms of scale, composition and changes in production technologies. However, eliminating export subsidies is part of the Doha agenda because of their price effect. By subsidizing production in the exporting country, these subsidies reduce the price in the importing country, sometimes below the cost of production. They also increase the use of fixed factors of production, discourage agricultural diversification, result in over-grazing and the extension of farming to marginal land, and lead to intensive use of chemicals (fertilizer and pesticides). By suppressing world prices, they reduce farm incomes in non-subsidized countries and stifle poverty-alleviation efforts in many developing countries.

Export subsidies can also affect environmental conditions in a commodity-importing country when they are removed, by spurring rapid growth and investment that is environmentally damaging. Removing them can also cause the relative sizes of different sectors to change because they have insulated certain sectors from market signals.

If polluting sectors grow when subsidies are removed, they can cause environmental damage. Removing subsidies can also contribute to changes in production methods that are less costly but more polluting. Export subsidies are expected by be eliminated by the Doha agreement by 2015, but the agreement will begin to rationalize production in response to market signals before then.

Competitive producers will benefit, but if market signals act to increase productive capacity, this could be at the expense of the environment in sensitive areas unless corresponding policy measures act to discourage production there.

Externalities

This heading includes policy measures that are external to the cost of producing oilseeds, but highly relevant to national policies that affect the environment. These include, enforcement of environmental regulations and well-defined property rights. Both can add to the cost and risk of production but are not directly included in the cost.

Relevant environmental regulations can include requirements to retire land because of soil erosion, air quality regulations pertaining to airborne dust, regulations prohibiting wetlands loss, requirements for water quality/ impairment from nutrients, regulations governing wildlife habitat loss, environmental liability regimes, etc. Obviously there is great opportunity for variations in enforcement of these regulations between countries, areas within countries, and kinds of producers.

It can generally be assumed that environmental performance standards are lower where enforcement of environmental regulations is lax or nonexistent, but there may be exceptions for some areas where private certification schemes can effectively take the place of state intervention. Environmental enforcement is generally more effective with respect to processing industries than production of agricultural commodities.

Undefined property rights and lack of environmental enforcement can also deter investment in a sector and ultimately lead to creation of environmental problems when land is abandoned or otherwise not used productively. It is well-accepted by economic and developmental experts that markets for environmental (and all other) goods fail to work well when property rights are undefined, are poorly defined or cost too much to enforce. Property rights issues arise in the oilseeds sector in countries where agricultural expansion is taking place. As such, they are implicitly distorting and can lead either to an inability to use available resources or to uncontrolled exploitation unregulated by rule of law.

Non-pollution Environmental Externalities

Non-pollution externalities are the physical conditions in producing countries that have an effect on their ability to realize gains in terms of market access, etc., from trade reforms focused on agricultural production or processing.

These include soil quality, existing biodiversity, and geographical conditions like access to ports. For example, a country may not be able to expand production to take advantage of increased market access if it has no land available. Conversely, it may be possible for producers to further exploit already-degraded, cheap and available land for quick gains enabled by new market access opportunities. Non-pollution spatial externalities like soil erosion and biodiversity loss are therefore implicitly distorting and can act as barriers or incentives to exports. The Doha agenda will not change any of these parameters, but a producing country's policy response to them will be influenced by the Doha results.

Pollution Externalities

Like non-pollution spatial externalities, pollution externalities can act in some cases as barriers and in other cases as incentives to unregulated production and exports, depending on the terms of trade. Some examples of pollution externalities relevant to oilseeds production might include chemical residues in soil or water, air pollution caused by surface transportation or polluting processing facilities, or water pollution caused by fertilizer runoff. Policies relevant to pollution externalities might include becoming a signatory to environmental treaties or initiating programs to combat adverse environmental conditions.

ENVIRONMENTAL EFFECTS

The rows of the matrix assess effects relative to a set of environmental indicators. Environmental indicators were much discussed in the 1990's. The OECD Ministers adopted a set of environmental indicators in May, 2001. Of the OECD indicators, which include a range of indicators from climate change to fishery resources, several are particularly relevant to environmental conditions prevailing in oilseed producing countries. These include soil erosion/organic content, groundwater contamination, loss of habitat/ biodiversity resulting from forest conversion or improper land use, loss of other productive use, and deleterious human health impacts. As relevant as these basic indicators are, each will not be equally relevant to oilseed production in each producing country. The OECD indicators do not include soil health, for instance, which is a key component of sustainable agricultural production. The World Bank and other development institutions have also used environmental indicators extensively in their evaluations of projects for environmental and social impacts. Most recently the 2005 Environmental Sustainability Index (ESI)30 ranked countries on criteria that include overall environmental stewardship in several categories. They include not only soil, air and water quality and quantity, but also include environmental governance, reducing ecosystem and population stress, reducing transboundary environmental stress, and a number of other criteria.

REFERENCES

Badami, R.C. and K.B. Patil. 1981. "Structure and occurrence of unusual fatty acids in minor seed oils". *Prog. Lipid Res.* 19:119–153.

Broun, P., S. Boddupalli, and C. Somerville. 1998. "A bifunctional oleate 12-hydroxylase: Desaturase from *Lesquerella fendleri*". *Plant J.* 13:201–210.

Cahoon, E.B., C.R. Dietrich, K. Meyer, H.G. Damude, J.M. Dyer, and A.J. Kinney. 2006. "Conjugated fatty acids accumulate to high levels in phospholipids of metabolically engineered soybean and *Arabidopsis* seeds." *Phytochemistry* 67:1166–1176.

Cardone, M., M. Mazzoncini, S. Menini, V. Rocco, A. Senatore, M. Seggiani, and S. Vitolo. 2003. "*Brassica carinata* as an alternative oil crop for the production of biodiesel in Italy: Agronomic evaluation, fuel production by transesterification and characterization". *Biomass Bioenergy* 25:623–636.

Drexler, H., P. Spiekermann, A. Meyer, F. Domergue, T. Zank, P. Sperling, A. Abbadi, and E. Heinz. 2003. "Metabolic engineering of fatty acids for breeding of new oilseed crops: Strategies, problems and first results". *J. Plant Physiol.* 160:779–802.

Grushcow, J. and M.A. Smith. 2006. Engineering high-performance biolubricants in crop plants. Indust. Biotechnol. 2:48–50

Kinney A.J. 1996. "Development of genetically engineered soybean oils for food applications". *J. Food Lipids* 3:273–292.

Murphy, D.J. 2002. "Biotechnology and the improvement of oil crops—genes, dreams and realities". *Phytochem. Rev.* 1:67–77.

Bibliography

Amiruddin, Mohd, N., "Palm oil products exports, prices and export duties: Malaysia and Indonesia compared," *Oil Palm Industry Economic Journal,* Vol. 3 No. 2: 21-31, 2003.

Badami, R.C. and K.B. Patil. 1981. "Structure and occurrence of unusual fatty acids in minor seed oils". *Prog. Lipid Res.* 19:119–153.

Basiron, Y., N. Balu, and D. Changrmohan., "Palm oil: the driving force of world oils and fats economy," *Oil Palm Industry Economic Journal,* Vol. 4 No. 1: 1-10, 2004.

Broun, P., S. Boddupalli, and C. Somerville. 1998. "A bifunctional oleate 12-hydroxylase: Desaturase from *Lesquerella fendleri*". *Plant J.* 13:201–210.

Cahoon, E.B., C.R. Dietrich, K. Meyer, H.G. Damude, J.M. Dyer, and A.J. Kinney. 2006. "Conjugated fatty acids accumulate to high levels in phospholipids of metabolically engineered soybean and *Arabidopsis* seeds." *Phytochemistry* 67:1166–1176.

Cardone, M., M. Mazzoncini, S. Menini, V. Rocco, A. Senatore, M. Seggiani, and S. Vitolo. 2003. "*Brassica carinata* as an alternative oil crop for the production of biodiesel in Italy: Agronomic evaluation, fuel production by transesterification and characterization". *Biomass Bioenergy* 25:623–636.

Casson, A., "Oil palm, soybeans and critical habitat loss," *A review prepared for the WWF Forest Conversion Initiative,* August, 2003.

Chand, Ramesh, 2004, "WTO and Oilseeds Sector: Challenges of Trade Liberalization", *Economic and Political Weekly,* February 7, 533-537.

Chand, Roja, *Agri Clinic: A boon to Farming, Agro India,* Vol.III, Issue 4, 2003, pp-12-14.

Dar, William D. (6 December 2007). "Research needed to cut risks to biofuel farmers" (in English), *Science and Development Network.* Retrieved on 2007-12-26.

David V. Alford, *Biocontrol of Oilseed Rape Pests,* 2003.

Dohlman, E., S. Persaud, and R. Landes, 2003, "India's Edible Oil Sector: Imports Fill Rising Demand", OCS-0903-01, *Electronic Outlook Report from the Economic Research Service,* USDA.

Drexler, H., P. Spiekermann, A. Meyer, F. Domergue, T. Zank, P. Sperling, A. Abbadi, and E. Heinz. 2003. "Metabolic engineering of fatty acids for breeding of new oilseed crops: Strategies, problems and first results". *J. Plant Physiol.* 160:779–802.

Gantzer, Hugh & Collean, "Mustard: The Versatile Spice," *Indian Spices,* Vol.35, No.2, 1998, pp.8-9.

Grushcow, J. and M.A. Smith. 2006. Engineering high-performance biolubricants in crop plants. Indust. Biotechnol. 2:48–50

Gulati, A., A. Sharma, and DS Kohli, "Self-Sufficiency and Allocative Efficiency: Case of Edible Oils," *Economic and Political Weekly,* 31(13): A14-A24, 1996.

Gupta, Shakuntala and Balwinder Singh, "Price Spread in Marketing of Groundnut and Rapeseed-Mustard in Punjab," *Indian Journal of Agril Mktg.*:12 (1 & 2), 1998, pp.128-136.

Herz, Jonathan, "Using and Maintaining the Ram Press," *Enterprise Works Worldwide,* Washington, DC, 42 p, 1997.

Julia, J.F., and D. Mariau. 1979. New research on the coconut mite *E. guerreronis* (K), in the Ivory Coast. Oléagineux (France) 34:181-189.

Keifer, H.H., E.W. Baker, T. Kono, M. Delfinado, and W.E. Styer. 1982. An illustrated guide to plant abnormalities caused by eriophyid mites in North America. U.S.D.A. Agric. Res. Services. Agriculture Handbook No. 573:5-121.

Kinney A.J. 1996. "Development of genetically engineered soybean oils for food applications". *J. Food Lipids* 3:273–292.

Kumar, Arbid and J.S.Chauhan, "Challenges in Rapeseed - Mustard Production," *SAARC Oils and Fats Today,* Vol.V, Issue 7, 2003, pp.32-33.

Labys, W. C. 1977. "Multicommodity Substitution Patterns in the International Fats and Oils Market." *European Review of Agricultural Economics,* Vol. 4, No. 1, pp. 75-84.

Landell Mills Commodities Studies, Inc. 1991. *Oils and Oilseeds: The Cost Competitiveness of Major Producing and Processing Countries.* New York, NY.

Medina-Gaud, S., and E. Abreu. 1986. "The coconut palm eriophyid mite", *E. guerreronis* (Keifer) in Puerto Rico (Acarina: Eriophyidae). J. Agric. Univ. P.R. 70:223-224.

Meilke, K.D. and M. Jay. 1997. *Zero-for-Zero and the Canadian Oilseed Complex.* A report prepared for Agriculture and Agri-Food Canada, Ottawa, Ontario.

Meilke, K.D. and M. Swidinsky. 1998. *An Evaluation of Oilseed Trade Liberalization.* A report prepared for Agriculture and Agri-Food Canada, Ottawa, Ontario.

Mohd, Vandan, "Marketing Extension and Commercial Farming," *Yojana*, July 1999, 1999, pp-37-39

Moore, D. 1986. "Bract arrangement in the coconut fruit in relation to attack by the Coconut Mite *E. guerreronis* Keifer". *Trop. Agric.* 63:285-288

Motey, Rashmi and Smita Lele, "Plastic Films for Processed Foods – Special Requirements," *Packaging India*, Vol.35, Issues No.-5, 2003, pp-19-31.

Murphy, D.J. 2002. "Biotechnology and the improvement of oil crops—genes, dreams and realities". *Phytochem. Rev.* 1:67–77.

Murthy, K.N., *Inter regional variations in consumption patterns in semi-arid tropical India: A demand system approach*, Indian Statistical Institute, Delhi Centre, Discussion paper 01-06, 2001.

Niesten, E.T, Rice, R.E., Ratay, S.M. and Paratore, K (eds.)., "Commodities and Conservation: The Need for Greater Habitat Protection in the Tropics," Washington D.C., Conservation International, 2005.

OECD. 1994. *The World Oilseed Market: Policy Impacts and Market Outlook.* Paris, France.

Pae, Peter. "Jatropha plant's oil studied as jet fuel - Los Angeles Times". www.latimes.com. Retrieved on 2008-06-05.

Persley, G.J. 1992. *Replanting the tree of life: Towards an international agenda for coconut palm research.* Redwood Press Ltd. Melksham 156 p.

Pope, Kevin, et.al., "Origin and Environmental Setting of Ancient Agriculture in the Lowlands of Mesoamerica," *Science*, 18 May 2001:Vol. 292. no. 5520, pp. 1370 - 1373.

Ram, Kewal, *Saarc Oils & Fats Today*, Vol – V, Issues – 11, 2003, pp-28-30.

Saharan, G.S., et al., *Diseases Of Oilseed Crops, Indus Publishing Company*, ISBN: 8173871760.

Shosteck, Robt. 1974. *Flowers and Plants. An International Lexicon with Biographical Notes.* Quadrangle/The New York Times Book Co. 329 pp.

Wood, Marcia. June 2002. "Sunflower Rubber?" *Agricultural Research.* USDA

Woodroof, J.G. 1970. *Coconuts: Production, processing, products.* The AVI Publishing Co., Inc. 241 p.

Young, C.E., and P.C. Westcott, 2000, "How Decoupled is U.S. Agricultural Support for Major Crops?", *American Journal of Agricultural Economics,* 82 pp. 762-767. 34

Index